The Invent T
Guide To More Fun

Josh Burker

Constructing Modern Knowledge Press

For information on direct or volume sales, contact the publisher:

Constructing Modern Knowledge Press

Torrance, California USA

CMKPress.com | sales@cmkpress.com

CRA043000 CRAFTS & HOBBIES / Crafts for Children

EDU029030 EDUCATION / Teaching Methods & Materials / Science & Technology

ISBN: 978-0-9994776-0-1

Series editor: Sylvia Libow Martinez

Cover design: Yvonne Martinez

Important note to readers:

Like anything fun, the projects in this book involve a risk of things going wrong. Please take into account your own and others skill levels, comfort with electronics or tools, and many other conditions. Different people will find different projects easy or difficult depending on many factors and experience. Do not undertake a project that makes you feel uncomfortable. The author and publisher cannot and do not accept any responsibility for any damages, injuries, or losses as a result of following the steps and information in this book. Always obey law, codes, policies, manufacturer's instructions, and observe safety precautions. Use common sense and have fun!

Contents

The Case for More Fun 1

Hard Fun 9

Making Making Work 13

Software 19

Hardware 23

Projects

Strange Controllers for Strange Projects 31

TurtleArt Cut Paper 39

Automata 47

Programmable Paper: Art with the Chibi Chip 57

Passive Amplifier for Mobile Phones 65

Light Up Jean Jacket 71

Beetle Blocks: 3D Turtle Geometry 77

Arduino Album Cover 91

Turtlestitch Embroidery 99

Multimedia Marble Machines 105

Cardboard Pinball Machines 115

Sidewalk Chalk Geometric Art 129

LightLogo 139

LEGO Gear Tinkering 155

Project	Software	Hardware	Supplies
Strange Controllers for Strange Projects	Scratch	Makey Makey	
TurtleArt Cut Paper	Turtle Art	Cameo Silhouette	Cardstock paper
Automata			
Programmable Paper: Art with the Chibi Chip	makecode. chibitronics. com	Chibi Chip Microcontroller	Copper tape, Chibi-tronics Circuit Stickers
Passive Amplifier for Mobile Phone			Insulating foam
Light Up Jean Jacket		LilyTiny Microcontroller	Conductive thread, LEDs, battery
Beetle Blocks: 3D Turtle Geometry	Beetle Blocks	3D printer	
Arduino Album Cover	Snap4Arduino	Arduino	Artist canvas
Turtlestitch Embroidery	Turtlestitch	Embroidery machine	
Multimedia Marble Machines	Scratch	Makey Makey, WeDo	Marbles, pegboard, copper tape
Cardboard Pinball Machines			
Sidewalk Chalk Geometric Art			
LightLogo	LightLogo	Arduino	Neopixel ring
LEGO Gear Tinkering		LEGO bricks and gears	

Note: All projects also use a variety of easy-to-find electronics, sewing supplies, recyclables, office supplies, and craft materials. Most use a computer. Some offer 3D print options for some supplies. See each project for a specific supply list.

The Case for More Fun

When I wrote The Invent to Learn Guide to Fun I hoped for, but did not fully anticipate, the popularity and widespread adoption of the book as a treasure trove of creative, whimsical technology and crafting projects. Almost daily I meet people new to maker education and project-based learning who find the book to be an invaluable guide to creating hard fun projects at home, in the classroom, library, makerspace, or any creative learning space. More experienced makers tell me they appreciate the book for the breadth of projects and the adaptability of each project. The projects have worked well, have been remixed in fun new ways, and have inspired many people to include making in their educational settings.

Two years later, I find that many people are still new to maker education and project-based learning in the classroom. As a faculty member at both Constructing Modern Knowledge and Design, Do, Discover, two extraordinary professional development workshops for educators, I see people who are sold on the idea of including crafting, tinkering, engineering, programming, and other maker skills in their learning spaces but are uncertain where to start or what to do. My goal for this book is be helpful in that quest, introducing techniques, skills, software, and hardware that they can teach others to use. Additionally, this book and *The Invent to Learn Guide to Fun* provide projects that people can get into with relatively few hurdles but that scale as one's skills increase. Both of my books are appropriate for makers of all skill levels because the projects are adaptable and remixable. I want the finished product that you work hard to build to be uniquely yours, customized as you see fit. The creative technologies and tools are powerful and adaptable to other projects once you know how to use them. However, just like in the first book, the technology is not at the forefront of the projects but rather is the "magic" that brings the craft to life. The technology "behind the curtain" powers our creations, animating them, lighting them up, and enchanting others.

People in maker education are always looking for guided explorations full of wonder and opportunities for creative self-expression. My Makey Makey Scratch Operation Game (scratch.mit.edu/projects/3137739/), in *The Invent to Learn Guide to Fun*, still gets upwards of ten remixes a week. Sometimes I can tell a new class is doing a deep-dive exploration of the Makey Makey and Scratch because a big group of users will suddenly remix the project! The combination of Makey Makey, cardboard, conductive material, and Scratch, as it turns

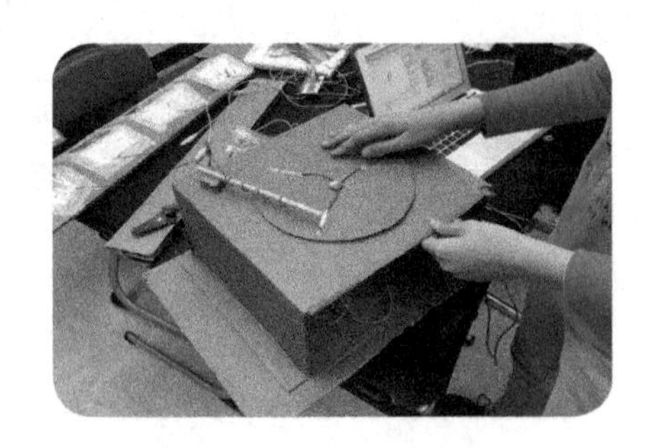

out, is a perfect example of Arthur C. Clarke's Third Law: "Any sufficiently advanced technology is indistinguishable from magic." How else can you explain the startling transformation of material headed to the recycling bin transformed into something as magical as a turntable that plays music snippets and small samples? You can see a video of the DJ playing her "wheel of cardboard" here: youtu.be/otoJyqNFIQM.

Makey Makey, Scratch, and the other technologies explained and demonstrated in this book are some of the tools that I use in my pursuit of outlandish "parlor tricks" that magically transform seemingly mundane materials. I call them parlor tricks because underneath the "wow factor" the mechanics or technology behind the feat is actually quite simple. One of my all time favorite parlor trick projects is a Makey Makey accordion. Jaymes Dec and I stayed up late into the night one summer at Constructing Modern Knowledge perfecting the "switches." Instead of relying on Scratch, we used a website (artpolikarpov.github.io/garmoshka/) that plays accordion polka songs when the browser window is resized! That's right, you click on the lower right corner of the browser window and drag the window left and right and the computer plays polka tunes!

Jaymes and I came up with a foolproof design that includes a little bit of a hack, too. After positioning the cursor in the lower right corner of the browser window with the "Garmoshka" page loaded, you use an alligator clip to connect the mouse click port on the Makey Makey to one of the Earth (ground) ports. This way the computer thinks you are clicking and holding. Meanwhile, the accordion, build around a three-part switch, one for mouse left, one for mouse right, and one for ground, is connected to the corresponding ports on the Makey Makey. When the accordion is "played" it causes the cursor to move left and right, and since the Makey Makey is sending a mouse clicked signal as well the browser window "magically" resizes as you ham it up with your cardboard accordion! I love this project so much I have built three different accordions over the years because it demonstrates the "magic" that simple materials and open, expressive technologies can conjure.

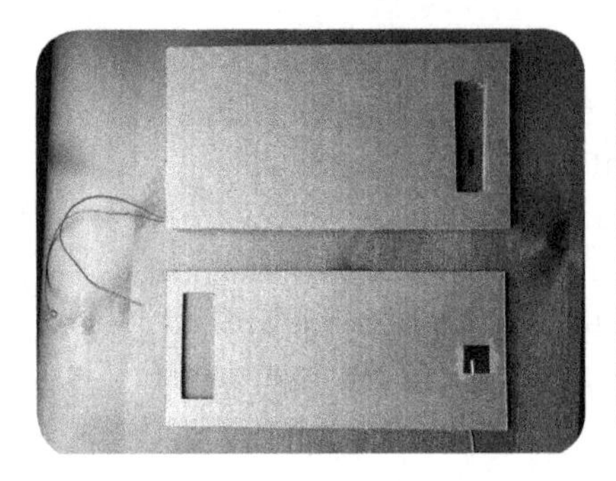

The success of *The Invent to Learn Guide to Fun* and (hopefully) *The Invent to Learn Guide to More Fun* in diverse educational settings as well as internationally (hello Canada, Ireland, Australia, New Zealand, the U.K., and everywhere else I have forgotten!) is in the customizable, remixable nature of the projects that encourage people to express themselves. Additionally, the low barrier to entry, whether in consumable materials, software, and to a large extent hardware, is very important to me because it helps to keep making inclusive rather than confined to private schools, well-financed makerspaces, or the rich white males who are the main audience of many "maker" experiences. Dr. Leah Buechley speaks more eloquently than I about lack of representation of women, makers of color, and other under-represented groups, and I encourage you to take the time to listen to her SITE17 talk about "Inclusive STEM Education": youtube.com/watch?v=KNcYWyqAll0.

Dr. Buechley and me

Making for All

I answered Dr. Buechley's call to "democratize" maker education after hearing about the outlandish cost just to join one of the robotics competitions popular in many schools. Rather than just gripe on Twitter about the problem of accessibility and the associated issues of race, class, and education in America, I did something about it. I collaborated with Brian Silverman and Erik Nauman to develop the LogoTurtle (joshburker.com/LogoTurtle/logoturtle.html), a low-cost microcontroller-based robot that many more people can afford to build, learn to program, and most importantly, debug.

I also worked with middle school students at a charter school in Bridgeport, Connecticut over the course of a school year to teach them Logo programming, fabrication, and

debugging. We used Turtle Blocks (turtle.sugarlabs.org) to create t-shirt iron-ons of our designs. Then we moved on to tuning the 3D printer that the school had received as a generous donation but were not using; 3D printed some of the students' Turtle Blocks designs; 3D printed the parts for a LogoTurtle; learned soldering and electronics to assemble to LogoTurtle; learned to successfully and diligently debug Logo procedures and create real knowledge about angles and degrees in the process. These students of color would otherwise not have had the opportunity to pursue such projects, but by leveraging what they had (a computer lab, a 3D printer nobody knew how to operate, and enthusiasm to build a robot) I was able to transform how they learned mathematics as we explored the beauty inherent in geometry.

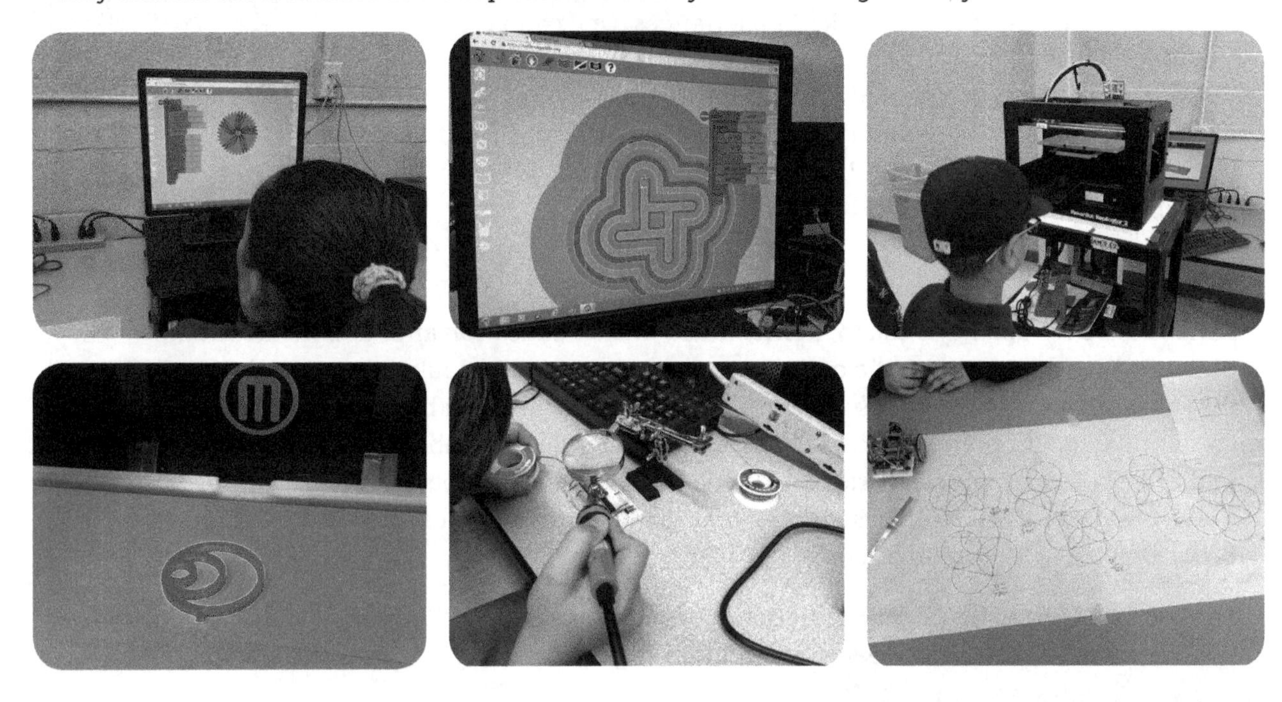

The project made such an impression on one student, who came to the after-school club the entire school year, that he asked to enter the LogoTurtle in the school's talent show! How often do you get to see a robot performing alongside amazing vocalists, dancers, and gymnasts? Isaiah brought the crowd to its feet! You can see his LogoTurtle's performance here: flic.kr/p/Rsg62r.

The success of this and other maker projects depends on de-emphasizing technology and emphasizing the crafting and tinkering (working with something to learn how it works) aspect of the project. The Logo Club was not about programming, but rather creating amazing, personalized and colorful designs of their own construction, first for the screen and then for clothing that advertised their programming skills without calling attention to the fact that the design was programmed. Their 3D prints were turned into functional jewelry. Even the multiple iterations of a design as the student debugged the design in its transformation from the screen to the LogoTurtle show how learning really happens. The march towards approximated perfection is a beautiful expression of their effort and tenacity.

You might think that if we spent more money on the robot it would produce more accurate results. While that might be true, it is not the point. These students took pride in the robot that they assembled themselves and even customized (notice the blue LED one student added to the LogoTurtle he and his peer made; the LEGO Minifig rider would be added later) and as a result took the time to debug the procedures, playing with angles and degrees, arcs, radii, and other mathematical concepts as they might a bin of LEGO or reproducing a route as they might read a map. These were not just math vocabulary words, but became a trusted part of the toolkit they used to make their robot, and make it do what they wanted. Providing students the opportunity to build and troubleshoot the tools that they can use to create their own knowledge about subjects that interest them is powerful learning in action.

Gender Inclusivity

Promoting collaborative opportunities and crafting possibilities in projects is particularly appealing to girls and women, who are typically underserved in technology-centric projects. The Marble Machines chapter in this book showcases a technology that surprised me when I brought it to the Danbury, Connecticut STEAM Fair. It was tinkered with for hours, primarily by women and girls! Since the success of any marble machine depends upon collaboration and patience, two skills girls develop sooner than boys, it makes sense that girls and women would enjoy playing with the marble machines. It was wonderful to watch their efforts be rewarded!

Lowering Barriers, Opening Doors

The LightLogo programming project explored in this book is a project that I combined with a crafting component to lower the barrier to entry. Sixth grade students collaged images representing their family stories onto the boxes that housed their LightLogo neopixel rings and Arduino microcontrollers. They programmed a version of the story they narrated on camera but told in light.

This project combined one of the more difficult programming challenges I offered to the sixth graders with a crafting project. Students who might not feel as strong at programming might better express themselves through their art or their writing, but all are given the opportunity to use technology. I saw them adapt their projects as their understanding allowed but everyone found success in all the components because of each component's low barrier to entry. The students' technology needs were supported by me, their writing needs by their English teacher, and their art needs by their art teacher. The students provided their teachers opportunities to collaborate, too, which made the lesson more inclusive and powerful in turn.

How to use this book

The projects in this book vary in the amount of involvement and the time it takes to complete a project. You can find a project to work on in the timeframe you have available. There are short projects that you can complete in an afternoon. Other projects are broken into manageable segments that can be worked on over the course of a school term or during a school vacation. Most of the projects work well for groups that can vary in size, and allow for a mix of abilities.

The materials used in the projects have a low barrier to entry, but with each project there is room to grow and take your creation to the next level. For example, your first iteration might be made of cardboard, but your final iteration may include 3D printed components.

The projects in *The Invent To Learn Guide to More Fun* encourage play and collaboration. You will find that nobody is sitting around when they are working on a project in this book—there are many roles to play, parts to build, and much fun to be had.

Like anything that's fun, the projects in this book involve a risk of things going wrong. Please take into account your own skill level, comfort with electronics or tools, and many other conditions. Different people will find different projects easy or difficult depending on unique factors and experience. Do not undertake a project that makes you feel uncomfortable, but don't always stay in your comfort zone. Always obey law, codes, policies, manufacturer's instructions, and observe safety precautions. Use common sense and have fun!

Each of the projects in *The Invent To Learn Guide to More Fun* is remixable, so make changes that personalize your project and make it your own. Share your remixes on social media and teach us a new way to play with the projects. Most of all, with *The Invent To Learn Guide to More Fun* in hand, go out there and have a good time!

Technology is always changing...

We have made every attempt at accuracy while writing this book. Technology will advance, products will be updated or even disappear, software will have new versions, and URLs are likely to break.

To save space and knowing that you cannot click on a book, we have shortened some very long URLs using bit.ly which will redirect you to the actual URL. We hope this website remains active!

Resources and URLs in this book, *The Invent to Learn Guide to More Fun*, and also the first volume, *The Invent to Learn Guide to Fun* can be found here: cmkpress.com/fun.

Hard Fun

I first became acquainted with Seymour Papert's concept of "hard fun" as a student of Dr. Gary Stager's in 2006. The ten "Learning Adventures" he challenged my Pepperdine University Online Master's in Educational Technology (OMET) cadre to take were vexing, outside of many of our fields of experience or expertise, and challenged us to collaborate despite physical distance from one another. Many of us had never played a musical instrument, but we were challenged to use Finale musical notation software to compose a musical piece. One Learning Adventure, where we were asked to examine the Collatz Conjecture (also called the 3n problem), was so difficult yet so engaging that I cornered a math teacher at a birthday party and talked with him for an hour about the problem and how best to analyze the data Dr. Stager taught us to generate.

I recall the vexation that some of my classmates and I felt with these learning challenges. The Learning Adventures had no definitive answers. Our anxiety to achieve in this graduate education course caused us to believe that Dr. Stager was somehow looking for a "correct" answer to the assignments. We were challenged to compose music, create podcasts (this was 2006, and most people had yet to start subscribing to podcasts, let alone creating them), and argue whether Ned Kelly (an Australian Robin Hood type) was a hero or a villain.

> To see how I used the MicroWorlds EX software to generate numbers and Microsoft Excel to graph the 3n data, visit cmkpress.com/fun/3n.

As hard as some of us searched, there was no "right" answer to any of these Learning Adventures. Rather, the exploration of a beautiful piece of astronomy software, like Celestia, or watching *Comedian*, the film about Jerry Seinfeld that Dr. Stager called the best example of a community of practice and of situated learning theories ever captured on film, provided opportunities to explore topics new to many of us. Exploring an idea without a notion of how one "solves" the problem is a powerful educational opportunity. Learners who are encouraged to explore and document how one approaches, turns over, and plays with the problem emerge from a Learning Adventure with a more flexible attitude to answering a challenge, solving a problem, or persuading others to a viewpoint.

> The OMET program is now named Master's in Learning Technology (MALT) and remains an intense, transformative one-year online Master's Degree in Education program offered by Pepperdine University.

Once I learned to embrace the adventure and find joy in being stymied because I was under no pressure to devise a "correct" solution, I began to understand what Gary meant when he kept mentioning "hard fun" as we complained about the difficulty of some of the Learning Adventures.

Figuring out that Dr. Stager wasn't after any one single answer to the Learning Adventures made the class a little more fun, but there was no getting around the fact the Learning Adventures were still hard. After all, each Learning Adventure seemed to bring us further outside our immediate areas of knowledge. We were teachers, business people, students, managers, and we were being challenged to create projects about history, the solar system, and computer programming.

At least with one project, where we were asked to use MicroWorlds EX to program turtles to draw quilt patterns, I had some immediate familiarity: I had briefly programmed Logo in elementary school, and my mother is a quilter. Still, this Learning Adventure required me and my cadre to learn (or re-learn) Logo programming. Dr. Stager has been instrumental in keeping Logo in the conversation about teaching children to program since the 1980s. In this whimsical challenge, we had to program not only our quilt square, but load our classmates' turtles into our MicroWorld and choreograph them to build an entire quilt from the turtles' procedures. Some of us had to dust off long-forgotten Logo programming skills, while others learned Logo for the first time. A true "low floor, high ceiling" computer programming language, most everyone was able to create unique, colorful, and collaborative turtle procedures. To make this project a little less hard and a little more fun, we needed to learn new tools. The success of a "hard fun" project depends on access to a large toolbox of tools, whether they are digital in this case, or physical, in the case of a 3D printing project. Success also depends on using the right tool for the right reasons for the right project.

As the Pepperdine course peaked and we were really rolling on the Learning Adventures, we found the secret to achieving the maximum "hard fun" from the Learning Adventures. Since the Pepperdine program was nearly entirely online, we used different tools, both synchronous and asynchronous, to keep in touch with one another and conduct our classes.

The more we collaborated, the richer the Learning Adventure became. Even though we were physically separated, we were programming turtles that performed in harmony, generating and sharing huge samples of data to explore Collatz's "3n Problem" looking for repeating patterns, and arguing over the martyr-hood of the Chicago 7.

Here is the explanation of how I designed my quilt square and programmed my classmates' turtles to create a quilt: cmkpress.com/fun/quilt

Participation and the exchange of hard ideas created the atmosphere of "hard fun" Gary kept promising us when we pushed back at the difficulty of the challenges. The authentic need to collaborate in these Learning Adventures created engagement and deeper understanding of the topic of the Adventure. Without collaboration, we would not have a pool of data or an audience to listen to a podcast about changing my diet and teaching you a new recipe.

If I distilled my understanding of "hard fun" into an easy to explain idea, it is an engaging open-ended project that pushes your skills to create knowledge that is shared.

This book is a collection of "hard fun" projects intended to help you find fun as you build skills, make things with your hands and the technology around you, and create learning environments where people collaborate around a project to further the design, use, and function of the tools you create. While some of the projects in this book can be built and used solitarily, they are most fun when brought into classrooms, conference anterooms, or your home, where many people can collaborate and express their creativity. Sharing the process and the tools reveals new uses and spawns remixes.

Making Making Work

There is no single magic recipe for "making" in education. But if you forced me to name one thing that is often overlooked, it's play. This may seem obvious in a book about fun, but we can create more powerful learning experiences for our students if we allow for play in the classroom, library, or makerspace. I am, of course, not referring to wild abandon, where people might get hurt or equipment damaged. Instead, I am speaking of the type of inquisitive, collaborative type of learning that occurs during play and to which the late Dr. Edith Ackermann drew our attention because of play's potential to help young and experienced alike activate new connections and knowledge.

Dr. Ackermann and me

In my keynote speech at the Pine Crest Innovation Institute 2017, I spoke about the role play should take in our learning spaces. This was one of the most fun talks to deliver because I let the students' work speak for itself. You can see the slides and videos here: bit.ly/morefungplay.

I defined play in the following ways:

- Play uses the body
- There is plasticity of identity in both objects and people during play
- Play involves taking things apart in order to understand them
- Play occurs in the world around us

If we consider the role of play in maker education, the connections are apparent. Maker projects naturally involve using the body: one crafts with one's hands, whether through handling tools or using a keyboard to program a computer or microcontroller. Projects like the Marble Machine, sidewalk chalk geometric art, or the cardboard pinball machine require you to use your body not only to assemble the project, but to use the project once constructed. Similarly, maker projects constantly repurpose, upcycle, and transform run of the mill items (I'm looking at you, cardboard) into extraordinary creations. Maker technologies like the Makey Makey, a literal "digital duct tape" that allows you to connect anything conductive to your computer to use as an input device, is transformative technology in literal and figurative senses. Whether one is debugging a program, disassembling an old DVD player in search of electronic parts to re-use in a project of one's own design, or trying to get a marble from the top of a pegboard to the bottom, much of maker education involves taking things apart. Finally, just as in play, making occurs in the world around us and is often a response or reaction to the world, the environment, or the needs of a particular community. The world provides context for the things people create and make, just as the world is the context in which play occurs. By leveraging these aspects of play and where they overlap with making and hard fun, one can nurture a natural aptitude for making in practically everyone and anyone.

Aesthetic Choices

In addition to emphasizing the role play serves in maker education, projects should also promote aesthetic choice. By providing familiar, easy to work with materials that people are used to using, designs can be customized and personalized to the maker's specifications. The connection forged between the student and the project in which she is personally invested because she is allowed to decorate, program, and present it in a way that she finds aesthetically pleasing is stronger than that made by a student churning out cookie-cutter projects that all look and function the same. Sometimes opportunities for aesthetic choice are as simple as providing paint to decorate the cardboard "unstruments" that you help your students build. Other times that choice becomes less obvious but equally important, such as allowing students to choose (or design) their own Sprite in a Scratch project rather than insisting everyone use Scratch the Cat or a similar character. Scratch as a platform values aesthetic choice, from its simple but powerful enough graphics editor, being able to upload your own sound files, to being able to remix, a very important possibility for learners today. How better to understand something than to be able to see its parts interacting and working? Scratch facilitates learning by allowing anyone to personalize a project, whether it is changing the gender of the hero Sprite or changing the code that makes the project operate as it does and making the project do something completely different.

Time to Debug

A vital way we can support makers is to provide them with the opportunity and time to debug, whether it is software, hardware, engineering, or other challenges impeding what the maker deems as being successful. Success often takes several iterations before it works consistently and accurately enough to want to share the work with others. We must provide learners with the support, in the form of encouragement, knowledge, access to knowledge if the problem exceeds our own abilities, and the time necessary to work through a problem. In a LEGO WeDo camp I ran for middle school students, we had access to a classroom with large joined tables for five days straight. Students were able to set up and leave in place their crazy engineering that composed a Chain Reaction machine. Perhaps most importantly, each work day came to an end and they took time away from the project, reconsidering any jams or difficulties they might have met that day and getting a fresh perspective before returning the next day with new approaches and solutions.

Emergent Design

This project touched on so many foundations of maker education that it is important to note them. First, the technology was just another tool that they used to accomplish the goal of triggering the next "machine" along the chain. We did not sit down and program the computers then try to fit the programs into the scene and machinery. Instead, the Scratch projects that helped power each machine evolved as the other materials came into place. Our sessions were filled with the quality play that encourages learning.

Students actively moved around the tables and rooms, helping one another and fine tuning the interactions between machines. They repurposed toys and parts to create new combinations and possibilities of movement and interaction, taking advantage of the plasticity of identity in play. They took apart systems and recombined them into the giant chain reaction machine, observing how the parts interacted, clashed, then finally meshed. Finally, we also created a world in which the materials were familiar or made familiar through use and tutorials, situating our play within the walls of the classroom and providing a context for our learning.

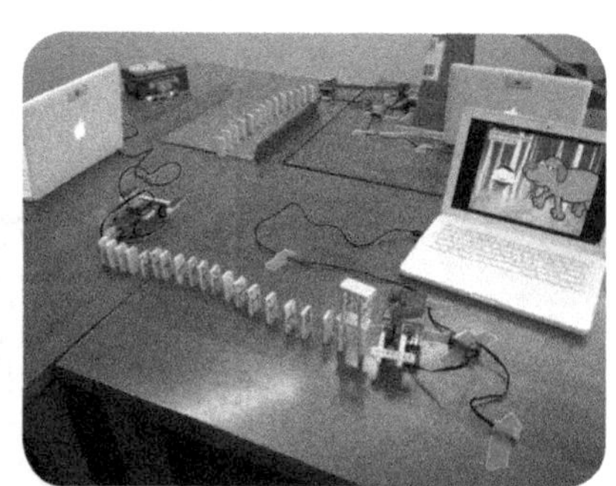

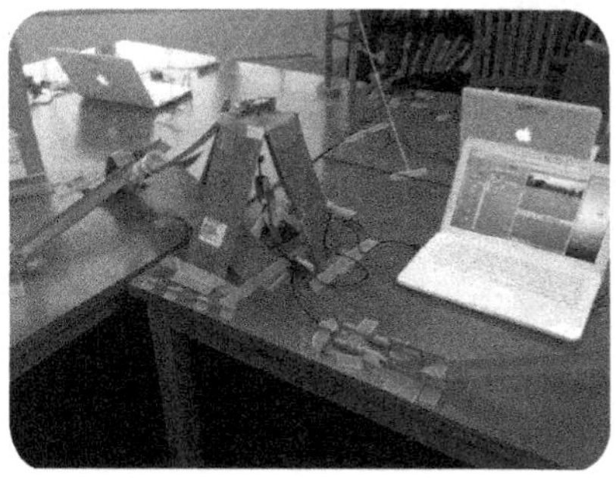

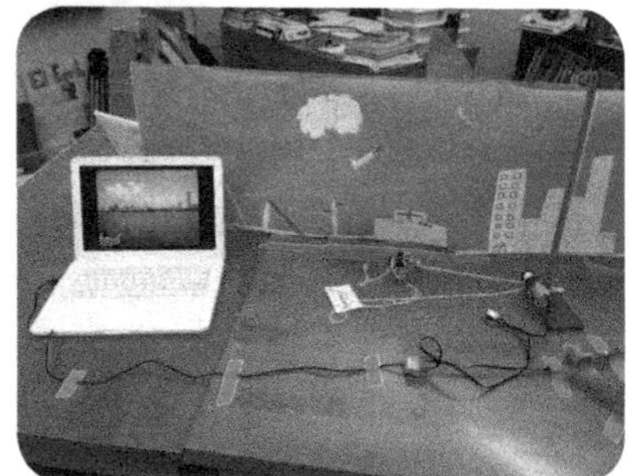

The materials, you will note from the photographs, included more than LEGO. The students used *bricolage*, or combining whatever materials they had on hand (and brought from home) to accomplish the job. Matchbox track combined with marble runs plus LEGO with a healthy helping of cardboard as well. The students aesthetic choices came out in the models they built and the way they decorated their particular scene. However customized their scene and machine, however, it needed to interact with those before and after it, so the success of the whole depended on the parts. People had to accommodate in their own design choices the way the whole model worked (both before and after their section) and everyone had to make compromises.

Collaboration

The collaborative nature of this project should be replicated in other maker projects. Collaboration brings out the hidden expertise that exists in any room. Encouraging your students to collaborate means you no longer need to be the expert in the room. As long as you, as the adult, can keep the room safe and are able to connect people with help, you are successfully promoting a collaborative makerspace. This help can be in the form of access to other people or information that you deem credible and helpful in the moment.

In addition to promoting a network within your learning environment, rely more on just in time learning rather than trying to teach every individual skill to everyone. This year's Maker Club, which I help facilitate with Dylan Ryder, uses a model of small groups working on two types of projects. Some projects are ones that students bring to the club such as a pom-pom rug for a faculty member's baby, or sanding the edges of a wine bottle he successfully cut at home and wanted to upcycle as a drinking glass. In other cases we provide open-ended prompts exploring tools and materials such as a marble machine construction with challenges of their own design or needle felting all sorts of cute and outlandish creatures and creations.

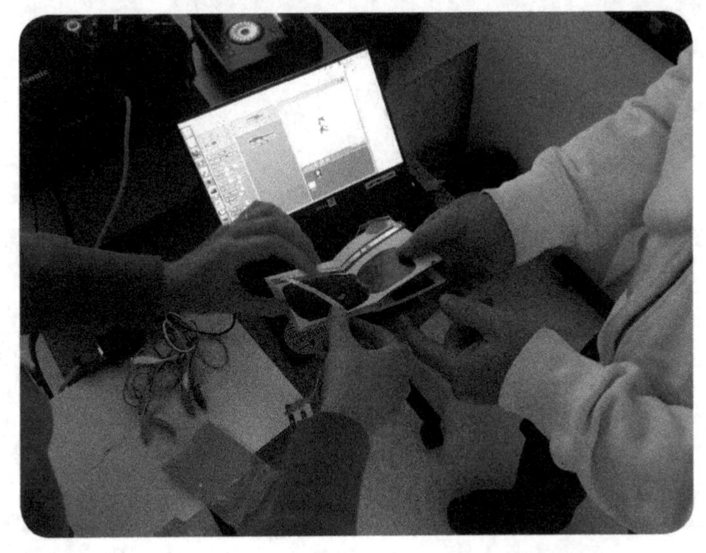

The personal investment in the project is important, as is the ability to work closely with peers interested in the same topic and working with the same tools and techniques to develop her or his work. I profess to know nothing about programming virtual worlds in the Unity Engine, but I helped students who wanted to install it on their laptops and let them support one another. I will continue to support them with access to information as they teach themselves and one another. These learning experiences, with peers engaged in projects that matter to those working on them, are the lasting lessons that affect who we are as makers and people.

Recipes or recommendations?

A final word about the structure of this book. Contained within this tome are both recipes and recommendations. This book strikes a balance between step-by-step directions that you follow to get a predictable result and product and snapshots of inspirational projects that share a technology but which are not explicitly reconstructed for you.

I mentioned the idea of tinkering parenthetically early in the first chapter, and I believe it applies to the recommendation parts of this book. I view tinkering as working with something to learn how something works. That something might be software, a 3D printer, conductive copper tape, or myriad components that comprise a marble machine. The recommendations are meant to show you the possibilities that I have encountered along the way. As I help others tinker with rich

materials, they will unleash the crazy invention they have trapped in their heads, perhaps not fully born but ready for a few iterations to arrive at the approximated perfection defined by the maker.

Both approaches of explicit "recipe" directions and tinkering "recommendations" speak to the need as maker educators to document the work of our students and ourselves and to share that work to inspire others. This book grew out of my obsessive documentation of projects in and outside of the classroom. I hope you and your students start documenting your work and sharing it, too, so others might stand on the shoulders of giants.

To that end, I hope you and your students, family, and others in your personal learning networks enjoy the projects in this book. I encourage you to share your work and your remixes on Twitter (you can @joshburker me on your tweets) and look forward to seeing the learning and hard fun happening!

Software

This chapter introduces the software used in the projects in this book. All the software is free and works on either the Mac or PC platform. There are many good tutorials and more information available online.

Scratch

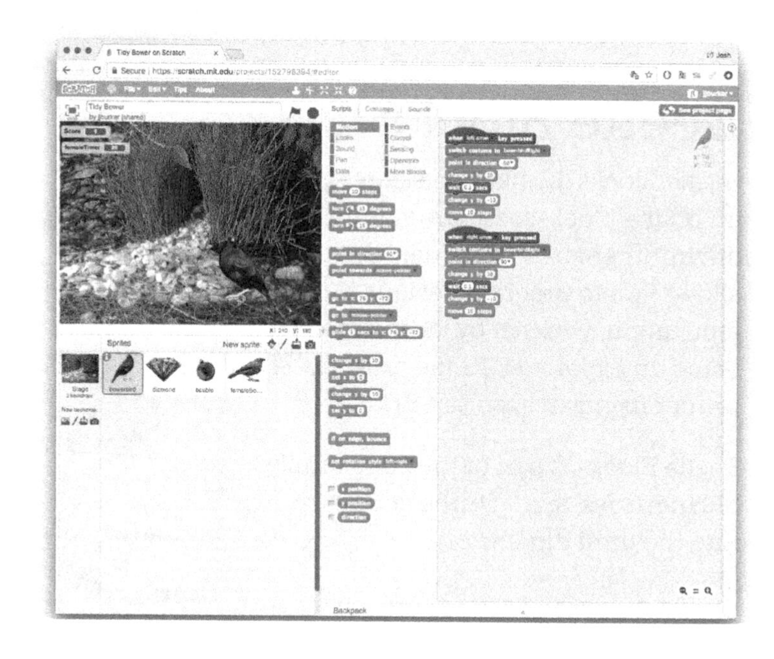

Scratch is a free programming language available from the MIT Media Lab's Lifelong Kindergarten group. Like TurtleArt, Scratch is a descendant of the Logo programming language that was invented by Dr. Seymour Papert, Cynthia Solomon, and others. These programming languages contain powerful ideas about how children learn that make them particularly relevant for today's classroom.

Scratch 2.0 (scratch.mit.edu) runs in most browsers and does not require downloading. An offline Scratch 2.0 editor is available for situations where internet is not available (scratch.mit.edu/download). Projects created in the Offline Editor can later be uploaded to the Scratch web site for others to play, examine, and remix.

Scratch 2.0 benefits from being cloud-based, so people who use Scratch 2.0 in a workshop go home with their projects online, ready to continue working on them. A lively, well-moderated user community exists as part of the Scratch community, where people can exchange programming methods, promote their work, and contribute to making Scratch better.

Scratch 2.0 also contains support for PicoBoards (sparkfun.com/products/11888) as well as LEGO WeDo 1 and 2 (see the Hardware section of this chapter for more information) to bring physical computing to your Scratch projects.

Snap4Arduino

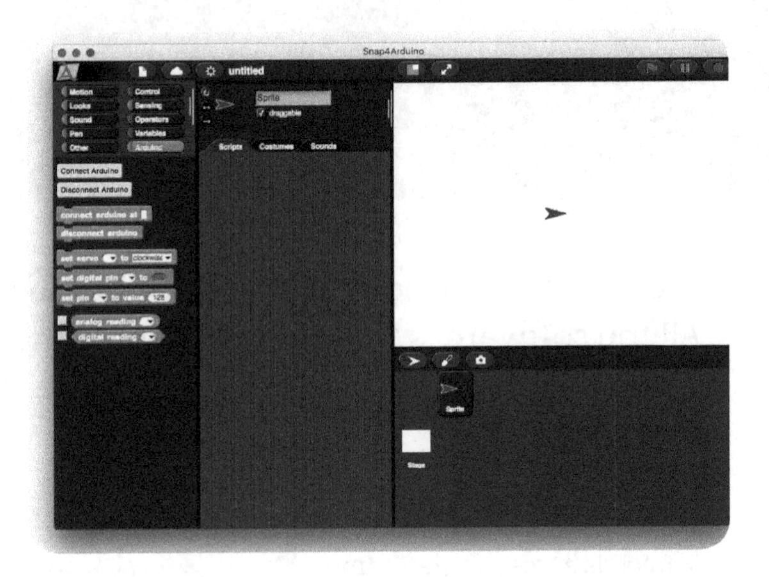

Snap4Arduino (snap4arduino.rocks) brings block-based programming to nearly every Arduino board. It lowers the floor to programming the Arduino because you do not need to type text and use the Arduino IDE syntax. Instead, you snap together blocks to control the Arduino and to collect data from it.

Beetle Blocks

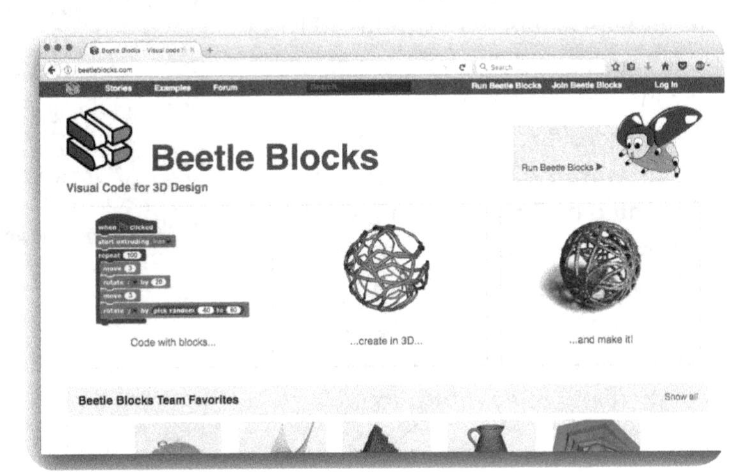

Beetle Blocks (beetleblocks.com) is an online block-based programming environment based on Snap! that allows you to program designs for fabrication, whether by laser cutter using an exported SVG file or with a 3D printer using an exported STL file.

Beetle Blocks is part of the Logo family of languages, taking turtle geometry into the third dimension.

LightLogo

LightLogo is a Logo programming environment written for the Arduino Uno or SparkFun RedBoard. It runs on a Mac or a PC. The included assembler.jar file allows you to load the firmware needed by LightLogo onto the Uno or RedBoard. There is good documentation included with the software as well as some fantastic samples programmed by Brian Silverman. (playfulinvention.com/lightlogo)

LightLogo

```
Welcome to LightLogo!
setc red
fd 8
setc green
fd 8
setc blue
fd 8
```

Tinkercad

Tinkercad (tinkercad.com) is an online 3D modeling Computer Aided Design (CAD) program that is usable by Kindergarten students and adults alike. Since being acquired by AutoDesk, they have continued to grow their connection with the maker education community, soliciting input and adding features, like class user management, to their product while keeping it free.

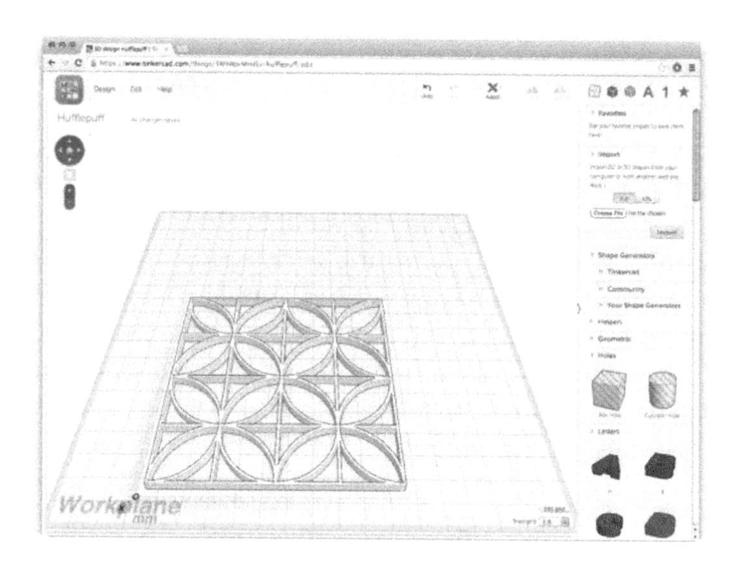

You can create 3D models using their pre-made geometric shapes, negative space, and javascript Shape Generators. Additionally, you can import .svg files to extrude and model for 3D printing. Tinkercad may seem simple, but its tools can be used to create incredible models. Its classroom features (like creating "classes" with student accounts) make it a great tool for makers of all ages.

Turtlestitch

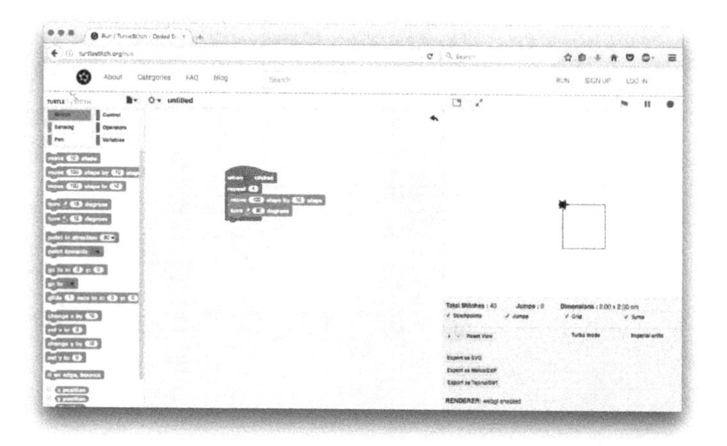

Turtlestitch (turtlestitch.org) is a block-based programming environment that produces designs and files for use with a CNC embroidery machine like the Brother PE-770. Based on Snap!, this program is easy to use. Since there are limitations to what an embroidery machine can physically embroider, determined by the hoop size, stitch length, and other factors, take the time to look at the Turtlestitch cards so your design can be realized in embroidery. (turtlestitch.org/static/download/TurtleStitch-Cards-Beginners.pdf)

TurtleArt

TurtleArt (turtleart.org) is an easy to use, block-based programming language that allows students to use mathematical reasoning, problem solving, counting, measurement, geometry and computer programming to create beautiful images. It is available by emailing contact@turtleart.org. Additionally, a version for the iPad, which takes beautiful advantage of the high resolution screen, is available from the Apple App Store. TurtleArt is part of the Logo programming family. At first, it may seem limited in its capabilities compared to other Logo implementations. However, it is easy to use and very powerful because of its elegant simplicity.

TurtleArt comes with a great two-part Getting Started tutorial and some samples and "snippets." Additionally, every image included in the TurtleArt galleries on the web site and the locally installed examples can be dragged and dropped into TurtleArt so you can examine the blocks and procedures. You can easily remix the sample projects, and people can easily remix your projects. If you post full size versions of your TurtleArt files, which save in PNG format, to image sharing websites such as Flickr, people can download the full sized version with the metadata containing the TurtleArt blocks intact for examining and remixing. TurtleArt also comes with a series of cards you can use to learn what the different blocks do and how to combine them to beautiful effect.

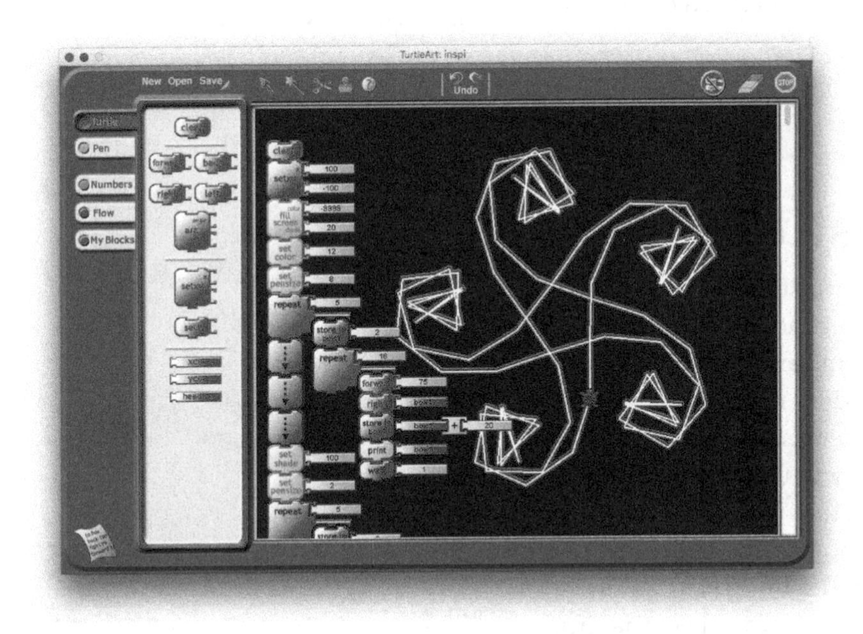

Hardware

All the projects in this book use various hardware, sometimes in combination. This chapter introduces the hardware and supplies you will need to complete these projects.

Makey Makey

The Makey Makey (makeymakey.com) is an interface board that at its simplest is a keyboard with the left, right, up, down, and spacebar. They include the mouse click for free, too.

The Makey Makey allows you to connect anything in the world that is conductive to your computer. When you hold the alligator clipped to the Earth, or ground, port on the Makey Makey and touch the conductive object you have also connected to the Makey Makey, the Makey Makey senses a closed circuit and passes on the key press to the computer to which the Makey Makey is connected. From there, anything you can do on your computer that uses those keys can be touch activated.

The possibilities with the Makey Makey are endless, and new inventions using the Makey Makey appear regularly. Once you become accustomed to the basics of the Makey Makey you can turn it over for access to additional keys as well as mouse movements and additional click functionality.

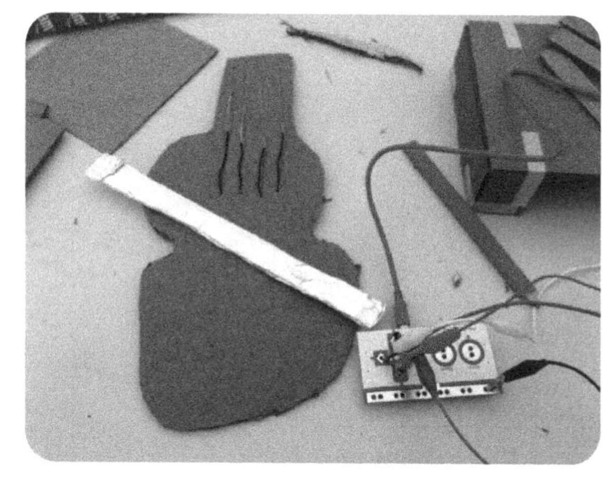

Arduino Uno or Sparkfun RedBoard

The Arduino Uno (store.arduino.cc/usa/arduino-uno-rev3) or the Sparkfun RedBoard (sparkfun.com/products/13975) are two versions of the same microcontroller. A microcontroller is basically a really smart switch that can be programmed to turn on and off LEDs, motors, or servos and gather data from attached sensors. Both the Uno or the RedBoard can be loaded with the LightLogo firmware that allows you to run LightLogo projects on them.

Brother PE-770

This CNC embroidery machine (bit.ly/morefunsewing) is a relatively low cost beginner's machine with many on-board options (fonts, borders, etc.). There are many YouTube tutorial videos about embroidery featuring this machine. Additionally, it reads DST files, which Turtlestitch (turtlestitch.org) produces.

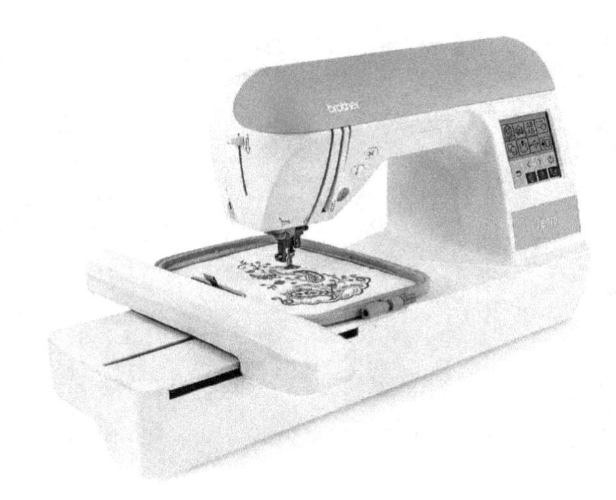

Chibi Chip

The Chibi Chip (chibitronics.com/shop/love-to-code-chibi-chip-cable) is a small form-factor microcontroller that you can program using a block-based editor (makecode.chibitronics.com) or a text-based editor (ltc.chibitronics.com). The Chibi Chip is thin and inexpensive enough that you can embed it in a paper circuit project. You download projects to the Chibi Chip using the included cable, which transfers the data via an audio signal. The Chibi Chip can be programmed from virtually any device, from a smartphone to a laptop computer.

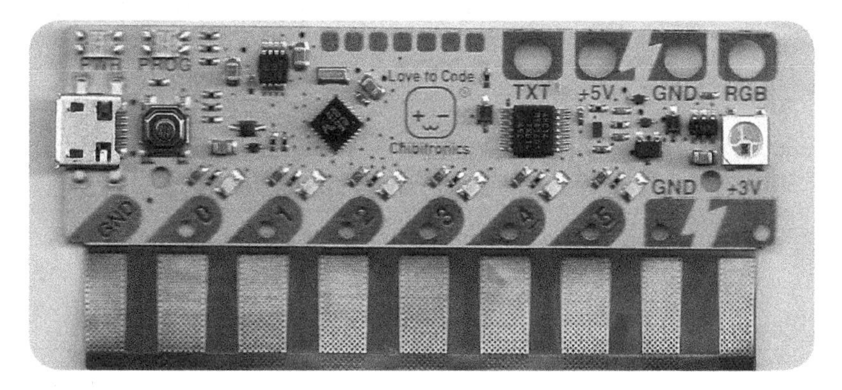

Silhouette Cameo 3

The Silhouette Cameo 3 (silhouetteamerica.com/shop/machines/cameo) allows you to cut paper and other materials from designs that you program or draw. The Cameo is very versatile, able to cut paper, vinyl, fabric, or cardstock. The included Silhouette Studio software (silhouetteamerica.com/software) is easy to use.

3D Printers

Access to 3D printers is becoming increasingly more common. Libraries, schools, maker spaces, and even individuals can take advantage of a variety of manufacturers as well as printing materials.

One consideration of 3D printing is the time the model takes to print. With a queue of work waiting to be printed, the task of producing a class or workshop's worth of 3D prints quickly becomes daunting.

Jaymes Dec (jaymesdec.com), who leads a Fab Lab at Marymount School in New York City, suggests that a few large format printers coupled with many relatively inexpensive, smaller footprint 3D printers that can produce smaller parts, is one solution to the problem of trying to produce many 3D prints. Another consideration is that perhaps not everybody gets a 3D print. Instead, people collaborate on smaller pieces that when combined form a larger object that can be displayed or placed in a school library for others to "check out" and play with themselves. A final option would be to utilize a service such as 3D Hubs (3dhubs.com) to have your models 3D printed for you.

LEGO WeDo

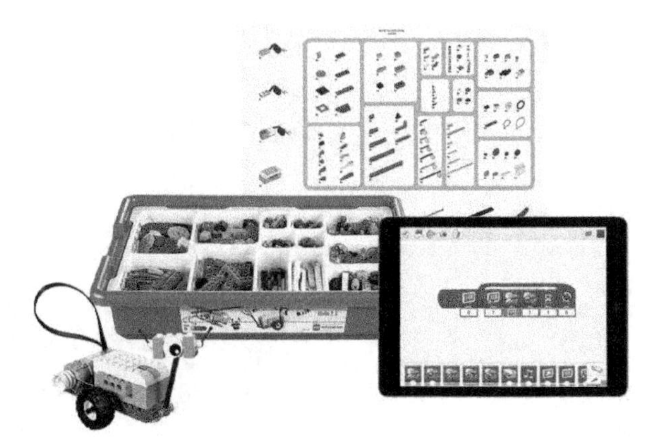

The LEGO Education WeDo 2 Core Set (education.lego.com) contains a Bluetooth Smart Hub, a tilt sensor, a distance sensor, a motor, and assorted bricks and connectors. It allows you to extend your existing LEGO bricks and sets by connecting to your computer and programming Scratch to read the sensors and turn the motor on and off with varying levels of power.

The original LEGO WeDo 1 Construction Set also works with Scratch. Projects that use WeDo 1 must be tethered to the computer via the included USB Hub.

NeoPixel Ring

The Adafruit NeoPixel Ring (adafruit.com/product/2862) is a 24 LED ring that can be connected to an Arduino Uno or RedBoard for use with LightLogo. Each LED can display red, blue, green, or white or any combination of those colors, making for nearly infinite lighting possibilities in your LightLogo projects. You will need to solder wires to the power, ground, and data connections on the NeoPixel ring.

Supplies

Each project in this book has its own list of supplies. Many are readily available around you in the form of recyclables or upcycled materials.

If you keep cardboard, wire, wire strippers/cutters, scissors, copper tape with conductive adhesive, pencils, and rulers on hand, you can easily start many different fun projects. Find new purposes in materials around you, approach your projects with a "make your own" mentality, and you will have fun and learn in the process. I hope these projects are inspiring for you and inspire you to remix, personalize, and expand upon them. Document and share your work and have fun!

Vendors

You can find many of the materials you need at local hardware and craft stores or even Amazon.com. The supplies list for the projects in this book suggest part numbers for some supplies, but you may find better (or cheaper) alternatives elsewhere. Some communities have recycling centers that cater to schools such as the Resource Area for Teaching (RAFT) network (raft.net). In addition, the vendors below carry many supplies, excellent learning resources, video tutorials, and best of all, many offer education discounts - be sure to ask!

- Sparkfun – sparkfun.com
- Adafruit – adafruit.com
- MakerShed – makershed.com
- Digi-Key – digikey.com

Projects

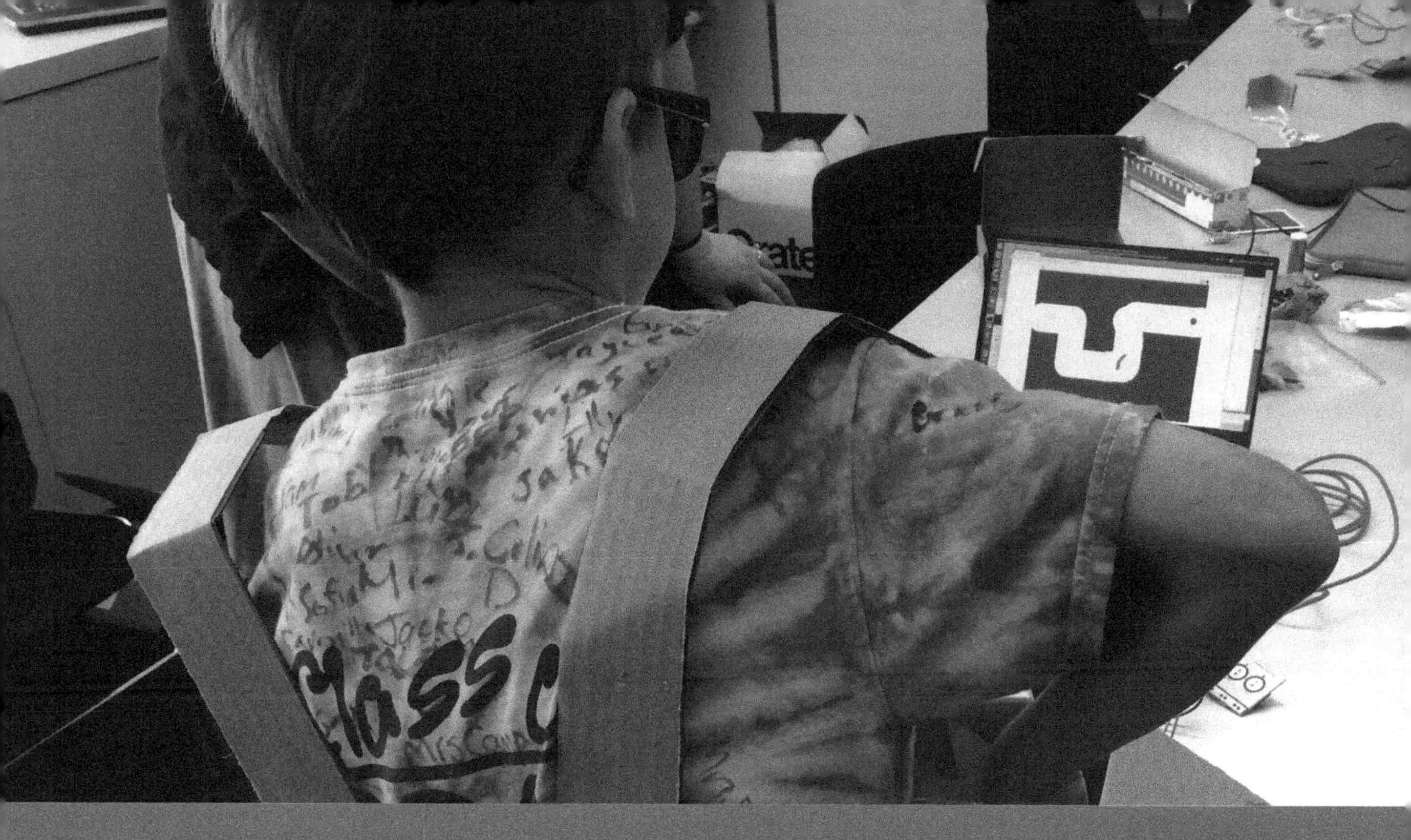

Strange Controllers for Strange Projects

This chapter showcases some of the very creative Makey Makey interfaces that various people, young and grown alike, have crafted into fantastic, imaginative controllers for their Scratch projects. Find inspiration in the ideas that these people brought to life and expand your imagination of what you, too, can craft with the Makey Makey, Scratch, cardboard, and conductive materials.

On the last day of a thirty day maker-in-residency, an 11 year old opened my eyes to even deeper creative potential in cardboard and copper tape than anyone, young or old, had previously. I helped about two hundred people create beautiful and oftentimes very accurate cardboard reproductions of musical instruments that used conductive copper tape and other materials connected to a Makey Makey and programmed in Scratch to mimic the sounds of cellos, violins, oboes, pianos, and drums. Blythe came to the last session with an idea already in mind to build a songbird. She quickly realized that she could bring her idea to life using the low floor materials of cardboard and copper tape plus the "magic" of programming in Scratch.

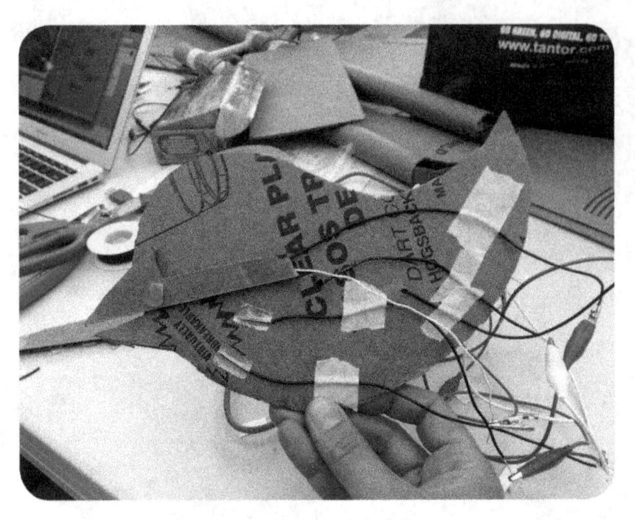

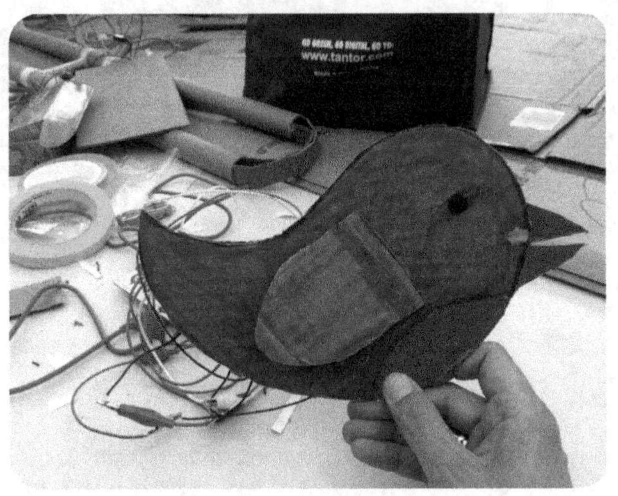

The "duct tape" quality of the Makey Makey brought together these two mediums to create her songbird. She learned how to turn the beak into a switch and how the switch could trigger Scratch to play different sounds. Not content with piano notes or bird sounds, she also learned on the spot how to use Audacity to pitch shift a bird song so she could play music according to a scale. She wrote sheet music for the song she composed and returned to the showcase to perform!

Rather than sticking to the obvious and building a traditional instrument, she chose to create something completely new. Projects like this are possible when you use an infinitely adaptable interface like the Makey Makey and construct with an equally flexible material like cardboard.

You can see a video of her songbird here: flic.kr/p/o8FYgA.

I have been fortunate to facilitate a few workshops where people create unusual controllers that used the Makey Makey and Scratch to play music, control games, and even teach us something new and fantastic. The hard fun of the Makey Makey and Scratch is turning a pile of cardboard and conductive materials like aluminum foil or copper tape into creative controllers for equally strange and outlandish Scratch projects.

Materials

- Makey Makey
- Computer running Scratch
- Alligator clips
- Small tools such as wire cutters/strippers, box cutters, scissors, etc.
- Cardboard boxes and tubes
- Conductive materials such as copper tape, aluminum foil, etc.

Make a Basic Instrument

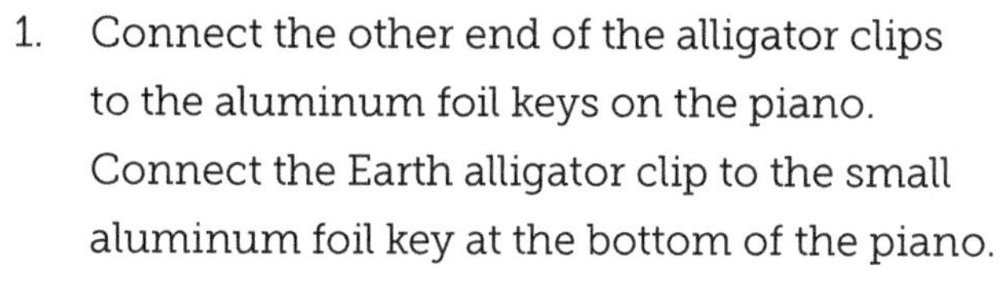

Makey Makey instruments all work primarily the same way. Create pads, keys, and touch points out of conductive materials, such as this piano. Connect the Makey Makey to the computer. Connect the alligator clips to the up, down, left, right, space bar, and the click ports on the Makey Makey. Connect an alligator clip to one of the ground, marked "Earth" on the Makey Makey ports.

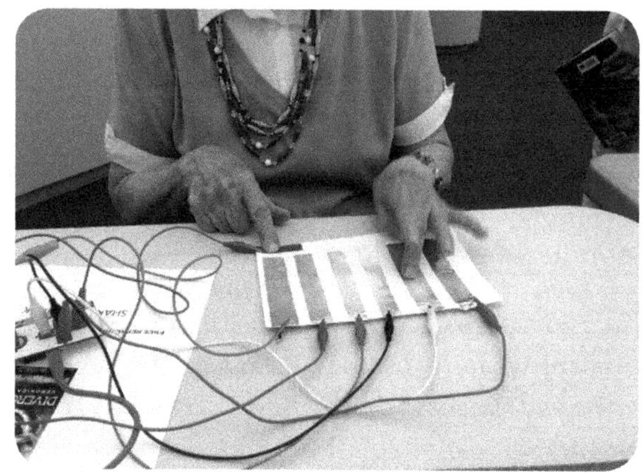

1. Connect the other end of the alligator clips to the aluminum foil keys on the piano. Connect the Earth alligator clip to the small aluminum foil key at the bottom of the piano.

2. On the computer, use Scratch to program the sound each key should make. Scratch includes a piano keyboard as part of the note selection block to help you easily program a piano keyboard. Scratch also offers other ways to play notes and sounds.

3. With one finger on the Earth key on your piano, you can touch the other keys to play the piano sound you assigned each key.

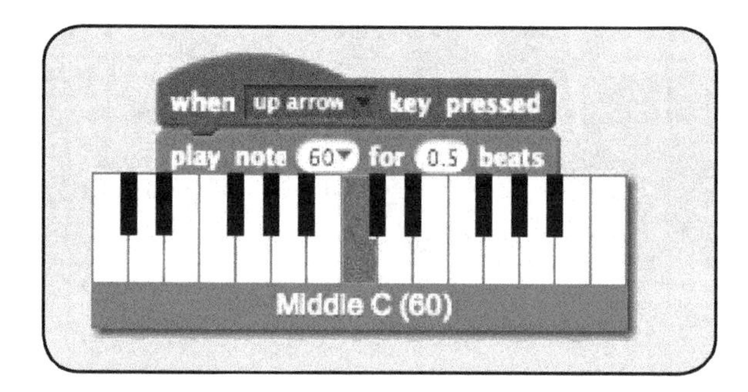

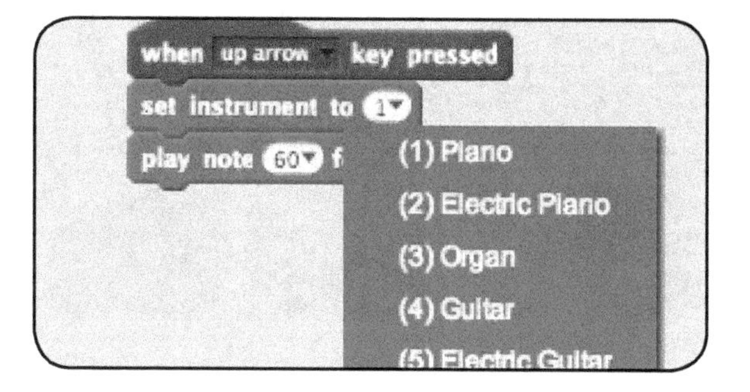

Strange Instruments

Since I love to use Scratch and MaKey Makey to create musical instruments, many of my workshops experiment with crafting creative instruments.

Unstruments

I often ask workshop participants to create musical instruments that don't look anything like the real instruments they sound like. Rather than listening to a lecture about taking risks, making a cardboard flute that sounds like a tuba is a signal that surprises are allowed and anything is possible. Prompts that encourage the unexpected and let imaginations soar open the door to creativity.

Monkey Instrument

This cute little monkey experiments with different conductive materials for its "keys." In addition to a large aluminum foil pad, it uses brass brads and conductive copper tape as well.

In addition to drum sounds, the monkey instrument chirps like a monkey and has a few other surprises. You can see it in action here: youtu.be/BOhe02dICKE.

Wearable Drums

This ambitious project turns a drum set into a wearable set of pads that can be programmed to play any sound you want. The mallet is connected to Earth on the Makey Makey, and when it touches the aluminum foil pads the circuit closes and Scratch plays a drum sound. This is a good example of using the Earth connection of the MaKey MaKey as the active element of the controller. A brief demonstration is here: youtu.be/1khtGugY3Xo.

Shark Instrument

This project transforms a large piece of cardboard into a denizen of the deep. The teeth are covered in conductive copper tape and act as the keys. The fin is the ground. At the culmination of the song we hear the terrifying shark attack! Watch the video: youtu.be/r_638g687-E.

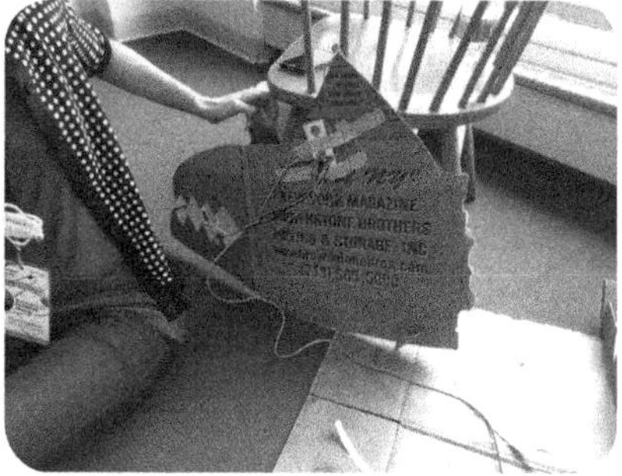

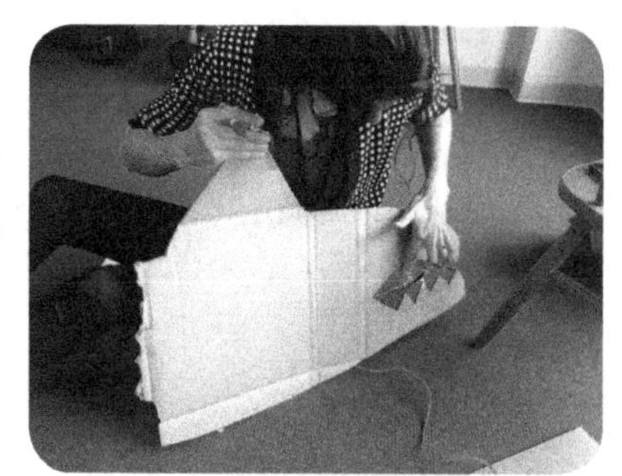

Flower Instrument

This whimsical instrument plays sound effects to accompany a song - all controlled by flower petals. Scratch can play multiple sounds at one time, so while one petal starts a song, the other petals can add sound effects in real time! The hollow flower pot base holds the Makey Makey. Unique, simple, and beautiful, this project needs to be seen and heard: youtu.be/IiKJk-b97nI.

The Beast

This crazy instrument was designed to play horror movie music! Crafted from cardboard and aluminum foil, this maker also composed music that was represented by colors that corresponded with the alligator clips attached to the instrument. Don't ask why it looks like two hockey sticks covered in foil, ask why not? You can hear it here: flic.kr/p/v74nxb.

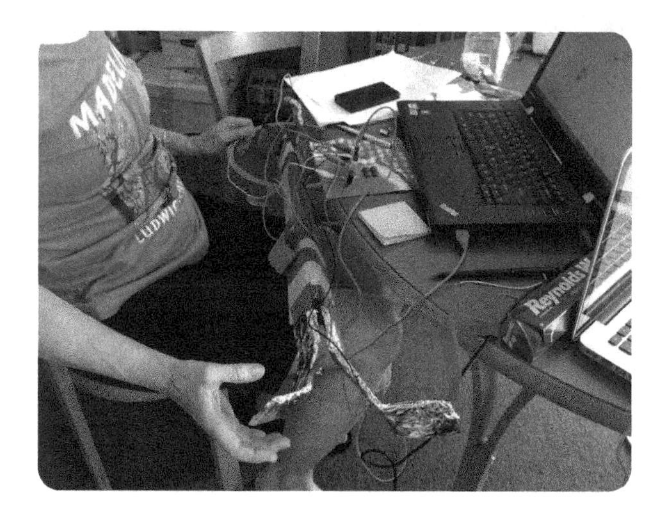

Strange Controllers

Music is not the only thing you can control with a Makey Makey and Scratch. Any game that responds to the Makey Makey can benefit from a strange controller—the stranger the better!

Earthworm Maze

This project grew out of a workshop for teens in which I asked them to come up with strange controllers for strange games. I ended up programming a simple maze game in Scratch where you play an earthworm. Add a strange controller and you get to BE the earthworm.

Scratch project: scratch.mit.edu/projects/70717770/

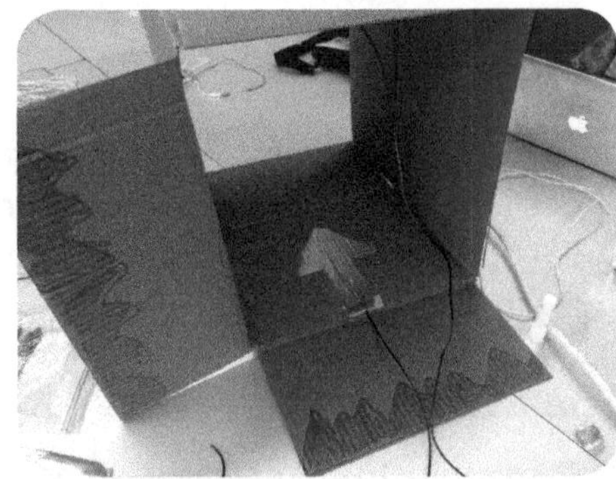

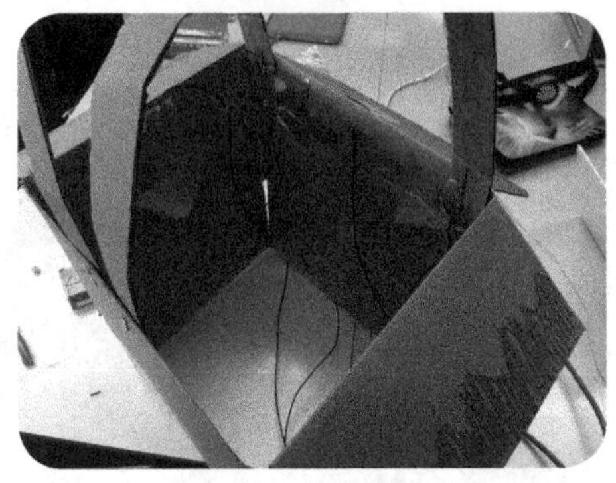

The wearable controller was built to look like a chunk of earth. The Earth pad is in the front of the box while the direction arrows are inside the box. While the maze is easy, it takes a little time getting used to the controls.

Tidy Bower and Invented Organisms Games

For a collaboration with seventh grade science I helped students build "arcade cabinets" that represented the habitats of invented organisms, remixing an idea from Dylan Ryder. As an example, I built one for a real organism, the bowerbird. Bowerbirds are known for building elaborate nests using a wide variety of found objects. Often, when I build examples for students, I change the prompt slightly so they don't feel that my example is the "correct answer."

The cardboard box fit around the laptop's screen so the Scratch project was visible in the back of the box. I added a piece of brass and a broken headphone as objects the bowerbird collected and which also acted as the controls to move the bird on the screen. Scratch project: scratch.mit.edu/projects/152798394/

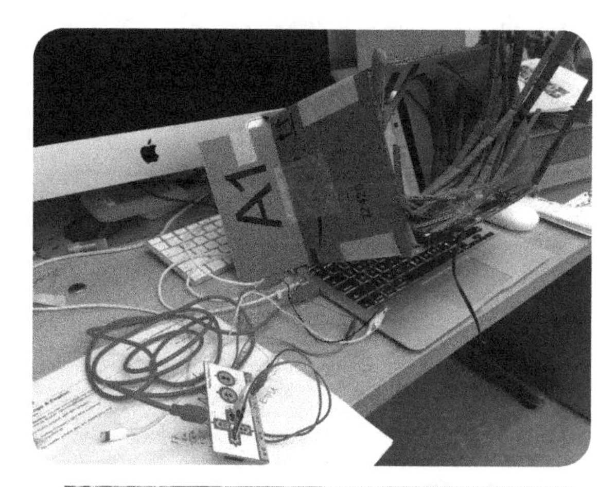

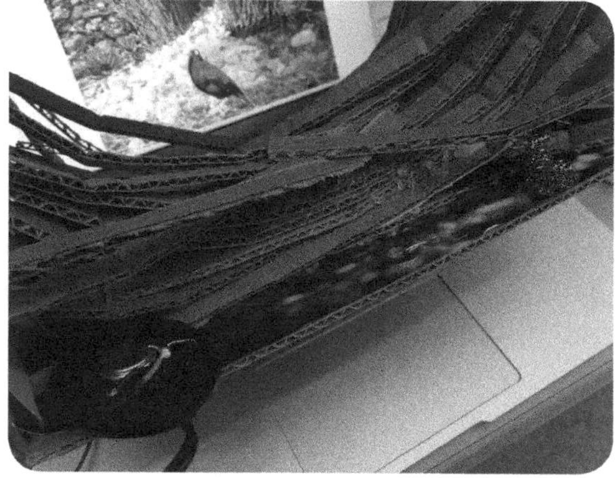

The students were wonderfully creative in building habitats in which their invented organisms lived.

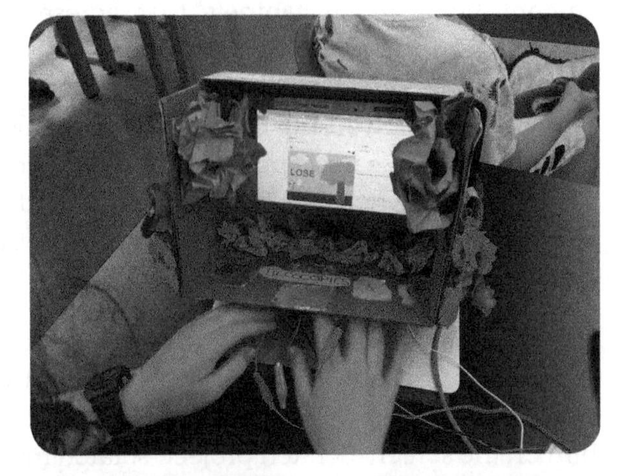

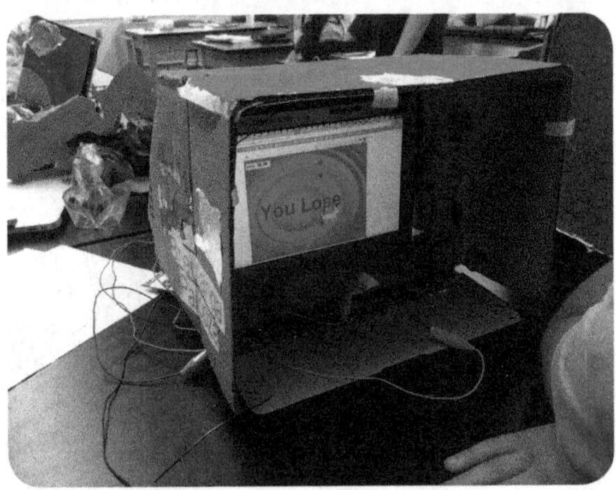

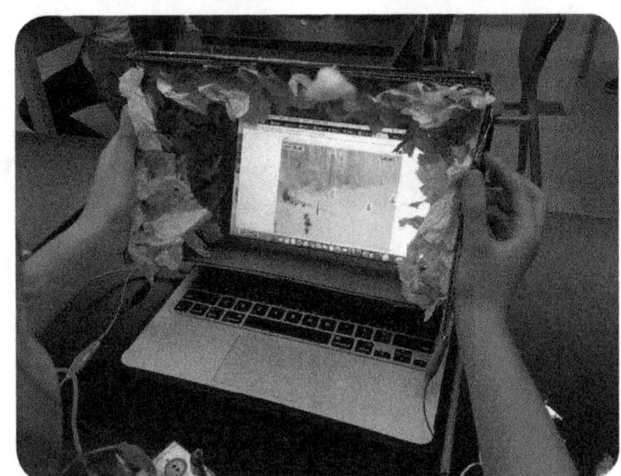

Taking It Further

Hopefully these creative builds inspire you to think big and go "beyond the banana" with your Makey Makey. Projects that combine basic circuitry and programming with a little bit of "strange" bring fun and magic to life.

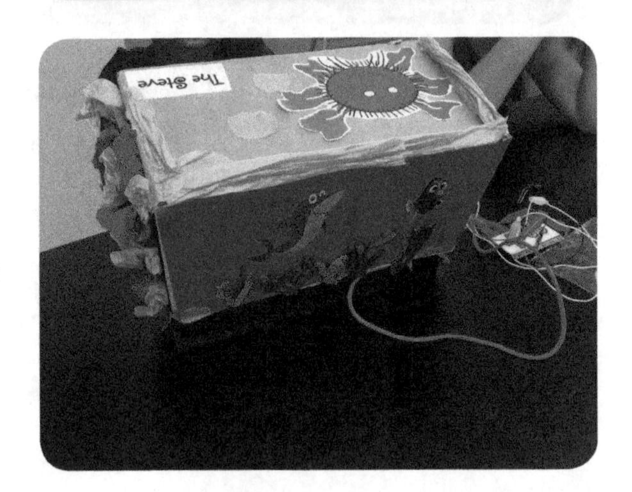

TurtleArt Cut Paper

The Cameo Silhouette offers a low-cost entry to bits to atoms fabrication, helping you to transform your digital designs into tangible objects made from cut paper or vinyl. Turning your TurtleArt designs into cut paper versions brings a new dimension (literally) to your work and broadens your creativity palette. Cutting your designs from paper is a remarkably easy task. The completed designs are beautiful on their own or can be incorporated into other projects with impressive results.

Materials

- Cameo Silhouette (silhouetteamerica.com/shop/machines/cameo)
- Cameo Silhouette cutting mat (silhouetteamerica.com/shop/blades-and-mats/CUT-MAT-12-3T)
- Silhouette Studio software, free version (silhouetteamerica.com/software)
- Computer to run the software
- TurtleArt (turtleart.org)
- Cardstock paper, 8.5 x 11 inches, in a variety of colors

Note: There are other cutting machines, but the Cameo is the one I use. If you shop for a machine, look for the ability to import designs, not just pull from a stock library.

Program Your Design

1. Program your design in TurtleArt. Since your design will be cut from paper try not to create too complex of a design, as the blade and paper may not be able to handle very thin lines, or too much material being cut from the paper.

2. Save your design. TurtleArt files are .png files, which can be opened by the Silhouette software.

3. Open Silhouette Studio. In the Design Page Settings, make sure you have the proper paper size and cut mat size set.

4. From the File menu, select Open, and select the TurtleArt design you programmed. Your design will open in Silhouette Studio.

5. You can resize your image to take up more of the paper. Click on the design, hold the Shift key on the keyboard and drag a corner to enlarge.

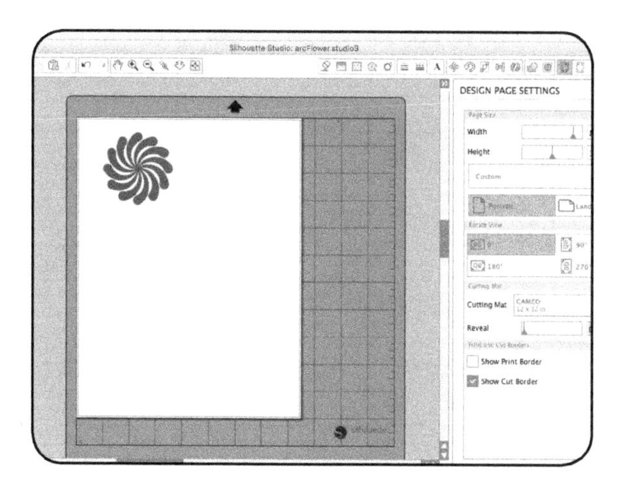

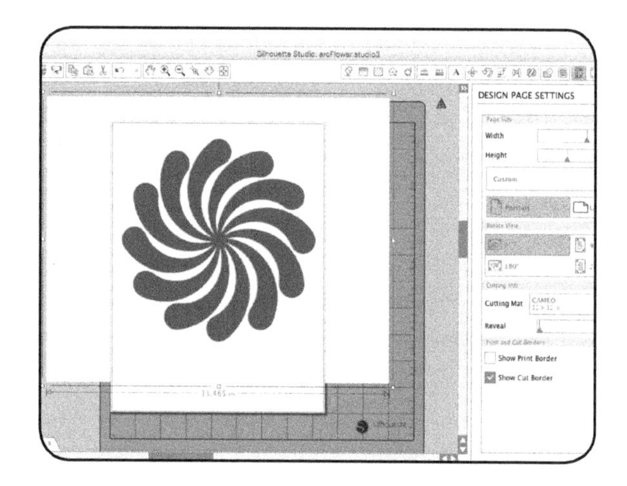

6. Click on the Open the Trace Window button in the toolbar.

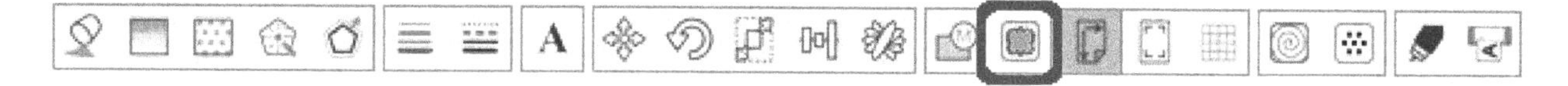

7. Click on the Select Trace Area button. Click and drag around your design.

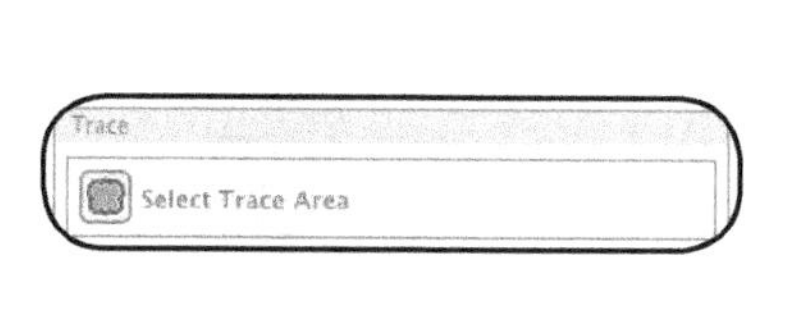

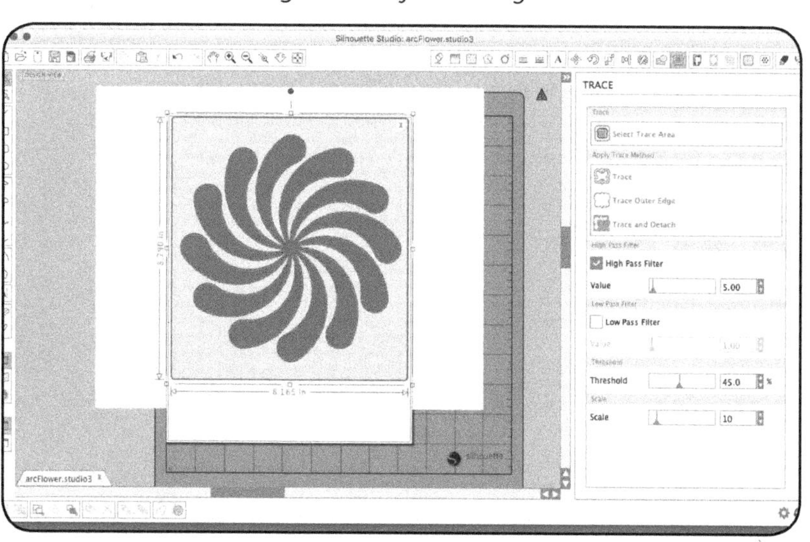

8. Use the High Pass Filter slider to make the design entirely yellow.

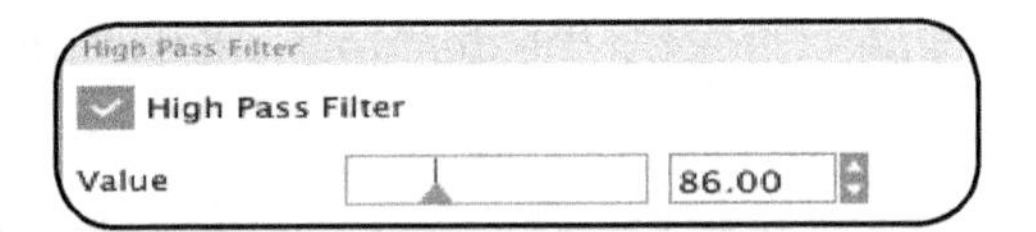

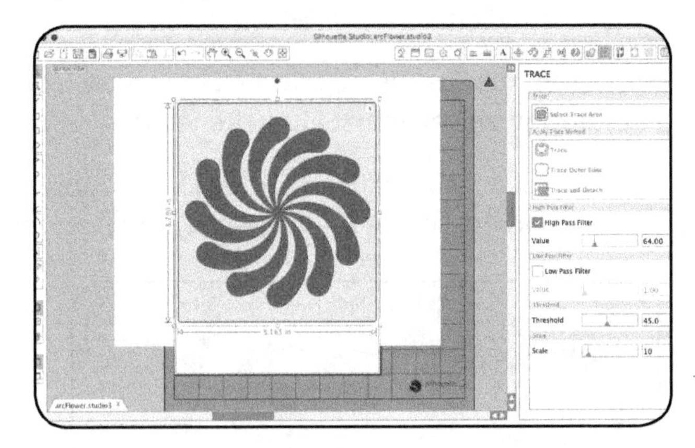

9. Click the Trace button under the Apply Trace Method options.

10. The design will be traced. You can move the original design to reveal the tracing underneath.

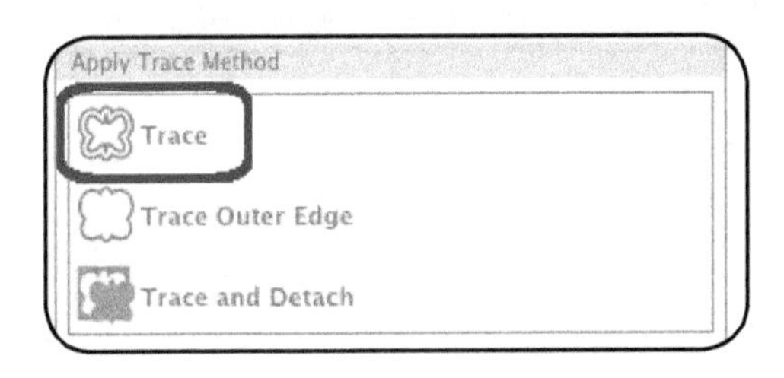

11. Press the Delete key on the keyboard to remove the original design, leaving the tracing behind.

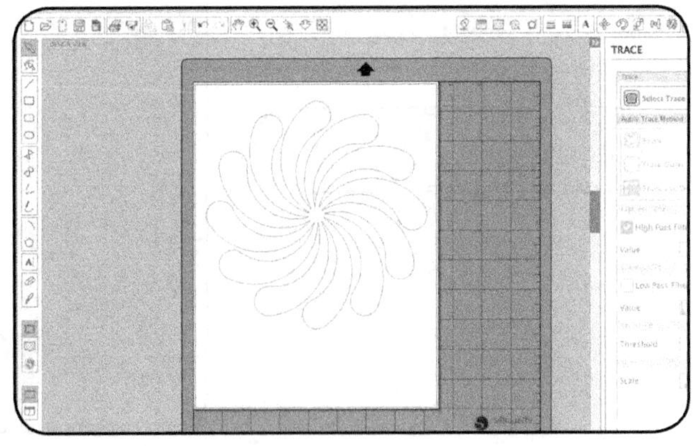

12. Click on the Cut Settings button in the toolbar.

13. Follow the manufacturer's directions to load the blade, properly apply the cardstock to the cut mat, and load the mat and paper into the Cameo.

14. Confirm that your settings match these settings.

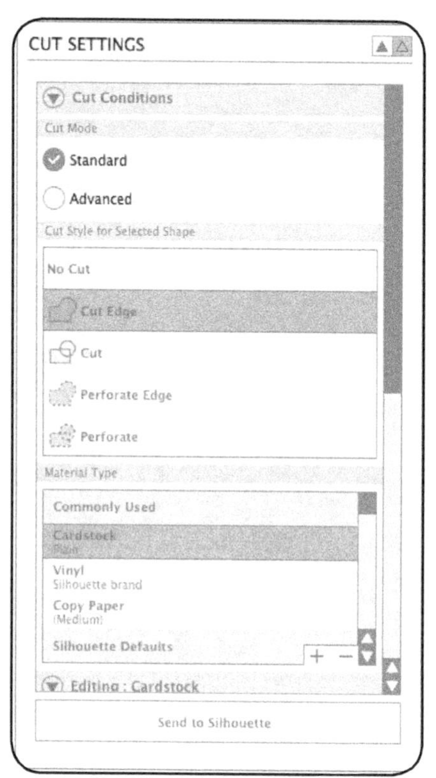

15. Click the Send to Silhouette button to start cutting.

16. Your design will be cut from the cardstock and unloaded from the Cameo.

17. If your design is particularly complex with many small areas or if your blade starts to get dull, you might have to remove some of the chads to clean up the design.

18. The variety of designs is nearly as limitless as those found on the screen.

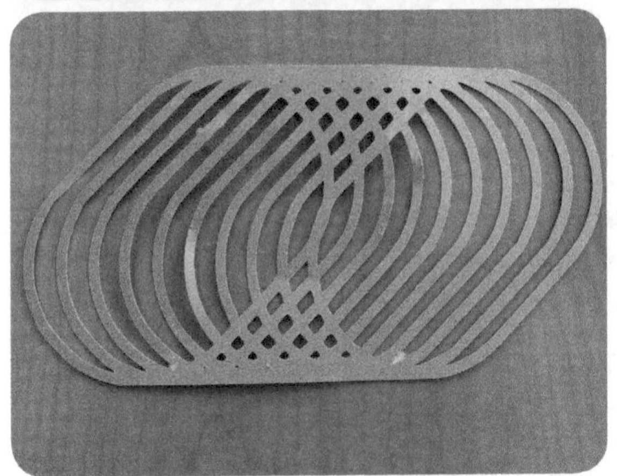

Variations

Some designs can be cut twice. Turn the second copy over to create a mirror version, and place on top.

Run the TurtleArt procedure in different pen sizes and cut each iteration. Stack the paper with small cardboard spacers between the layers.

It is fun to incorporate paper circuits and LEDs into the designs, too. See *The Invent to Learn Guide to Fun* to learn how to build illuminated paper circuits.

This student incorporated a 3D printed rivet (thingiverse.com/thing:551396) glued to the back of her designs. The rivet went through a hole in a small sheet of cardboard and attached to a handle in the back. This way the student could animate her design by turning the handle (flic.kr/p/SSuTec).

Taking it Further

Experiment with cutting vinyl in place of cardstock. You can create awesome car decals or laptop stickers. Incorporate your cut paper TurtleArt designs into other work. Decoupage them onto furniture. The world looks better with TurtleArt in it!

Automata

Automata are machines built to move in pre-programmed ways. Unlike computers that rely on digital programs or software, automata use mechanical (hardware) means to move. Using simple mechanisms, automata can seem to come to life, telling stories with their motion.

Automata have a long and colorful history that can be shared with students or workshop participants. Artisans combined complex gearing and clockwork to make amazing representations of humans, animals, and stories. Kings and queens were entertained by mechanical birds that flapped their wings and sang complex songs. Human-looking automata painted pictures, wrote letters, and played musical instruments, attracting large crowds of paying admirers as they toured the world. A chess-playing automaton called "The Mechanical Turk" was world famous until it was revealed to be a cheat—there was a secret compartment where a human chess player was hidden while he controlled the mechanical robot above.

Machines that seem alive still amuse and amaze us. This chapter explores some of the simple internal mechanisms that can be harnessed to create our own automata.

Telling a Story

A good automaton rewards its user by telling a short story with its movement. I built a series of automata that use nursery rhymes as their stories. Since most people are familiar with these stories there is an immediate emotional connection with the automata, and each provokes happiness in its user as she or he turns the hand crank and sets them in motion.

A good place to start with constructing automata is with the Exploratorium's guide, available at exploratorium.edu/pie/downloads/Cardboard_Automata.pdf. In addition to a list of the materials you will need to build your automaton, the guide includes illustrations of the different combinations of cams and followers needed to produce five different kinds of motion.

This chapter will lead you through some of my resources and showcase a few of the automata I built. You will find instruction and inspiration so you, too, can start creating mechanical devices that are engaging and fun to interact with and tell whimsical stories through their movement.

archive.monograph.io/joshburker/cthulhu-automata

Building Automata

You can build a simple automaton with items you probably have around your home or classroom. If you have never built an automata before, I recommend this project as a great introduction to automata for yourself or a workshop.

Basic Cardboard Box Automaton

1. Find the center of the box on both the face and the top of the box and mark it. Use a sharp nail to puncture a hole in the top of the box and through the center of the face of the box, all the way through, and out the other face of the box.

2. Use a sharpened pencil to widen the hole on the top of the box as well as the two holes in the faces of the box.

3. Cut a wire coat hanger long enough that it will pass through the hole in the face of the box and out the back with a bent handle on one end.

4. Use something circular (like the cap of a tube of hand lotion, like this student used) to create an oblong cam out of sturdy cardboard. Trace half the circle, move the cap over a couple of inches, and trace the other half of the circle to draw a cam in the shape shown below.

5. Use a nail to puncture a small hole in the center of the cam. Hold the cam inside the box then insert the coat hanger through the hole in the box face, through the hole in the cam, and out the other hole in the box face.

6. Now you will make a second cam called a "cam follower" that will sit on top of your first cam and transfer the motion to an object on top of the box. A cam follower can be cut from several pieces of cardboard and hot glued together. Cut two strips of cardboard and glue them on the bottom of the cam follower with a gap that the cam can fit into. These strips will act like guardrails to keep the cam follower from spinning on top of the cam when the cam is turned. A bamboo skewer with the pointed ends cut off can be inserted into holes you puncture in the center of the cam follower, as shown.

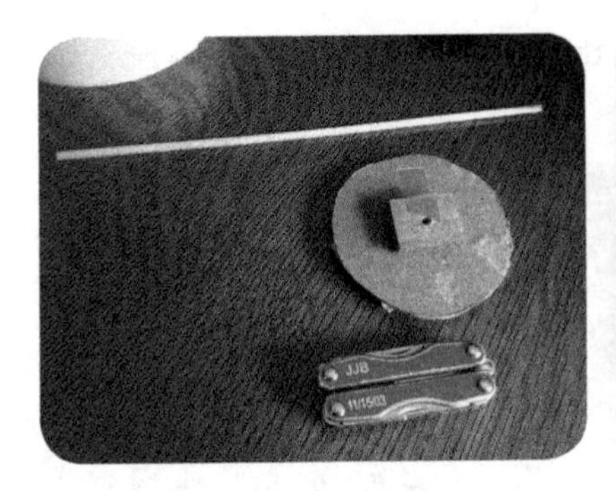

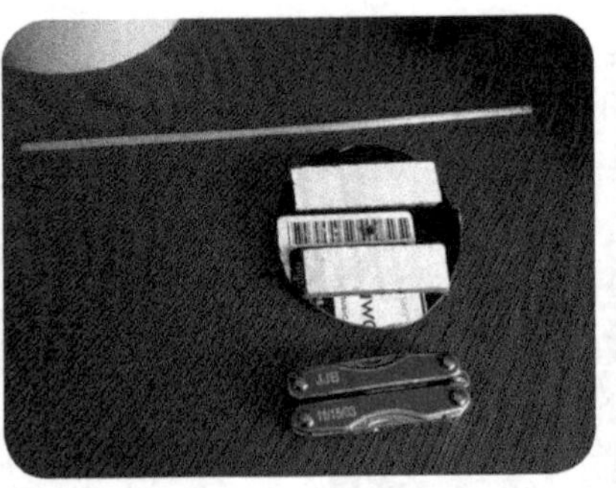

7. Cut a short piece of drinking straw to fit into the hole in the top of the box.. If it fits too loosely, wrap it in masking tape. Insert the straw in the hole in the top of the box.

8. Insert the cam follower in the box above the cam. The bamboo skewer will go through the drinking straw at the top of the box. Use a piece of masking tape to secure the cam to the crank so the cam spins with the crank when it is turned.

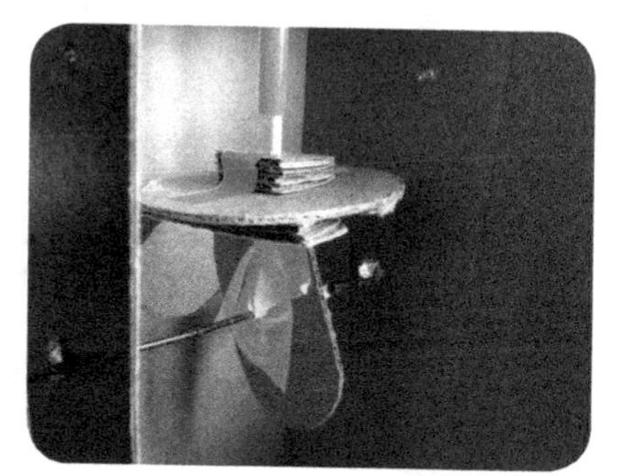

Observe the movement of your mechanism and then decide what story you can tell around that movement. This automaton moves up and down without spinning, but modifying the cam follower slightly allows the follower to rotate, too. After observing the movement, add something to the top of the bamboo skewer. Here I have added a little "person" to the automaton.

A cardboard automaton is a great place to start when first building automaton. The materials are easy to work with and you can experiment with different size and shape cams and cam followers.

Basic Automata Cam and Movement Types

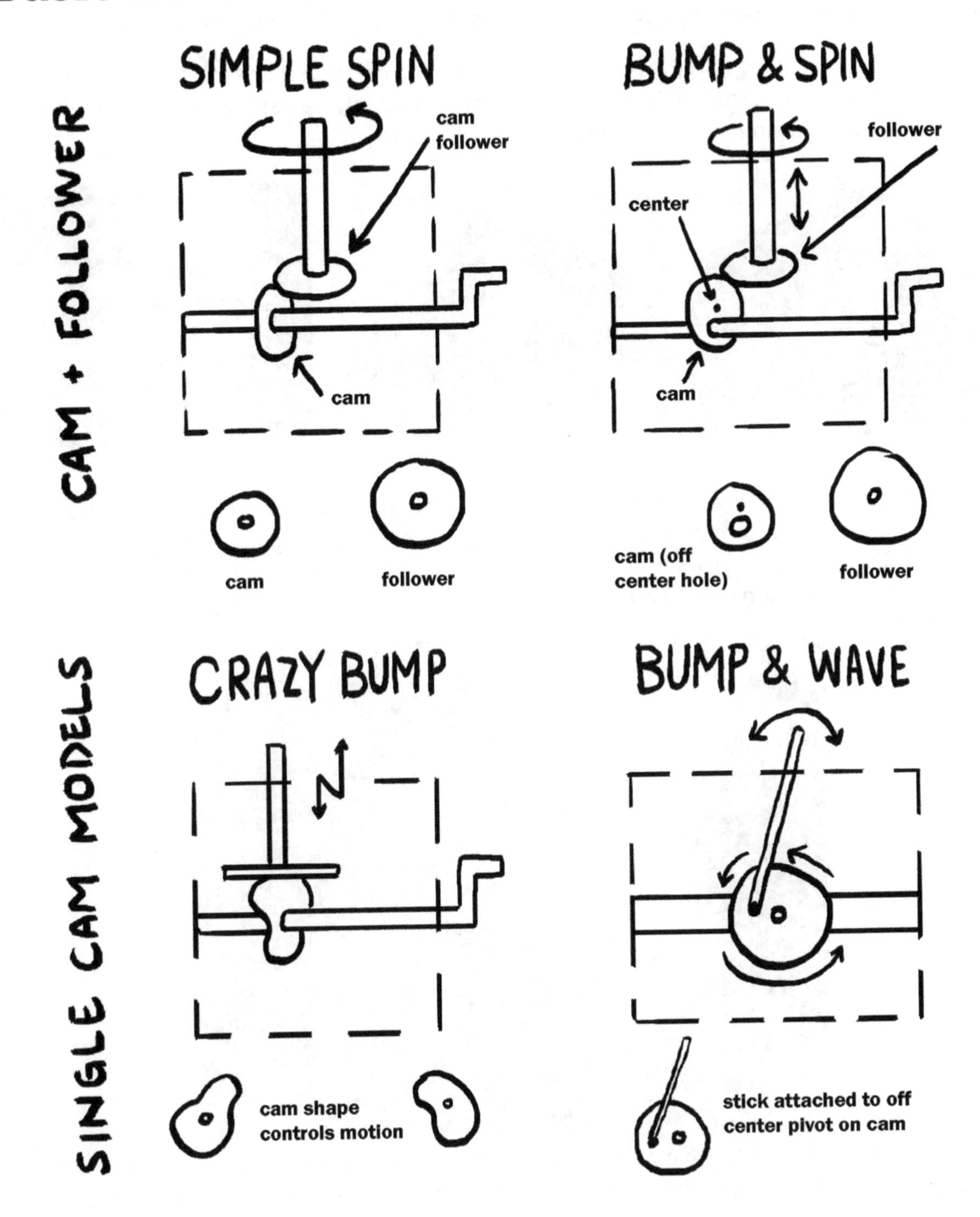

The Cow Jumped Over the Moon

This automaton started as a cardboard prototype with one storyline, but as it came together, changed to an entirely different narrative. It produces two different motions. One cam spins around, while the other cam moves back and forth in an arc parallel to the viewer.

Using the same cam layout and the dimensions from the cardboard prototype, I proceeded to craft a more durable version using a cigar box, wood dowels, and 3D printed parts. Instead of a crazed Boston terrier chasing a ball, this final automaton animated the fabled cow jumping over the moon in "Hey Diddle Diddle."

You can see a short video of this automata moving here: bit.ly/morefuncow. More images are available at this site: archive.monograph.io/joshburker/the-cow-jumped-over-the-moon.

Prototype

Final (front)

Final (front open)

Many of these automata were all built using remixed models that I found on Thingiverse (thingiverse.com) as the characters. I imported the models into Tinkercad to add a square hole for the dowels. This is a quick and effective way of creating the narrative elements of your automata if you have a 3D printer.

Hickory Dickory Dock

The "Hickory Dickory Dock" automata explored a variety of mechanical connections including cams and customized 3D printed gears created in Tinkercad to tell the story of the mouse running up the clock, complete with a ringing bell and spinning clock hands!

A short video of this automata in motion is here: bit.ly/morefunhickory. More images and video can be seen here: archive.monograph.io/joshburker/hickory-dickory-dock.

Tortoise and Hare

"Tortoise and Hare" is my latest automata. Again, I built it around a cigar box. This time, however, to add to the narrative I decoupaged the box with TurtleArt that I programmed.

View the classic race between the tortoise and hare here: bit.ly/morefunhare. More images and backstory are here: joshburker.blogspot.com/2017/01/tortoise-and-hare-automaton.html.

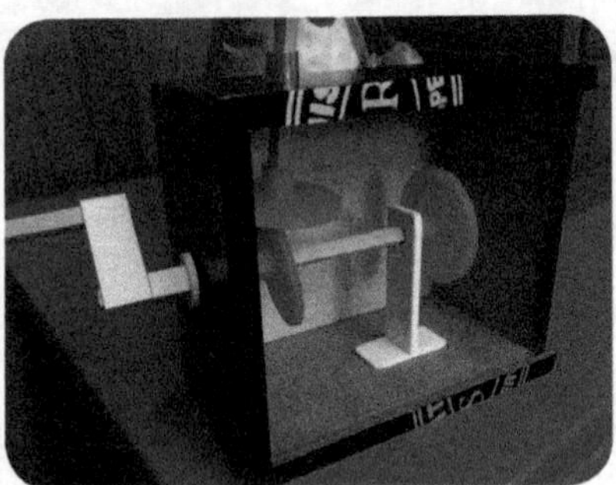

3D Printed Parts

In order to help you make more durable automata, I made available my 3D printable automata parts developed over the course of the construction of many automata.

These parts fit on a 1/4" square wood dowel. The square shape is important because the cams and followers will not spin on the dowel when put under load as you turn the mechanism.

You can download or remix these parts from Tinkercad: bit.ly/morefunprint.

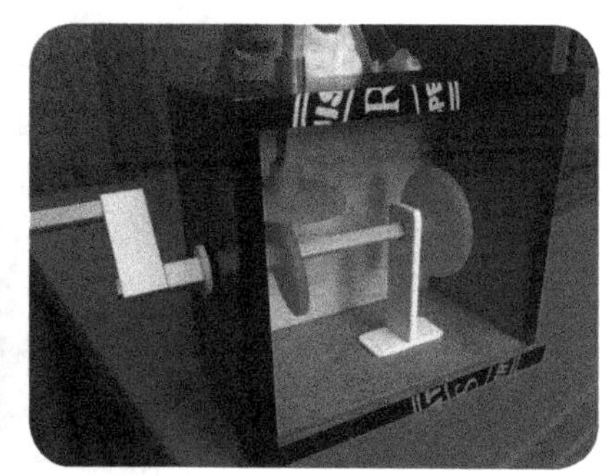

Taking It Further

For nine months my friend Joseph Schott, five students ranging in age from fourth to tenth grade, and I labored over the creation of a series of automata. We started with cardboard and learned woodworking and design skills. The students scaled their designs to four feet tall, two feet wide automata that told the story of the earth and the stars and organisms that live on our planet, ranging from starlings to a sperm whale chasing a giant squid.

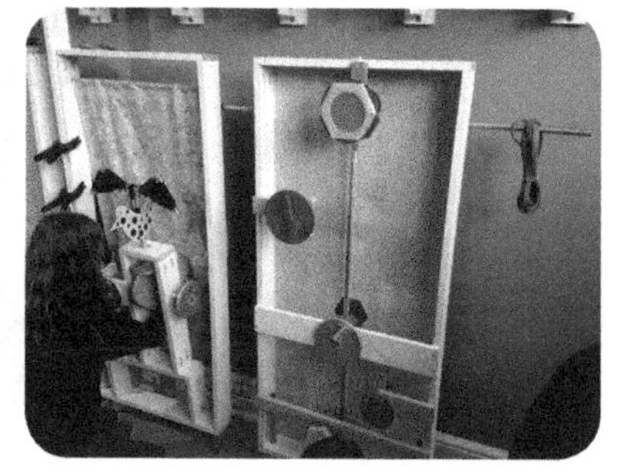

The five automata were displayed at the Westport Mini Maker Faire for visitors to interact with. These large automata use the same building techniques as the small ones, just with sturdier materials.

Now that you know how to construct an automaton and perhaps have found inspiration in my models, it is time to make your own. Make us laugh with your whimsical mechanical models!

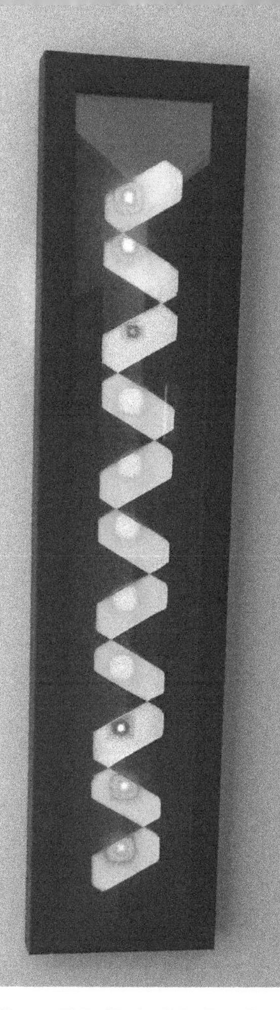

Programmable Paper: Art with the Chibi Chip

The Chibi Chip is a low cost microcontroller with a very slim form factor, making it ideal for embedding in paper circuit projects. The chip can be programmed to light up, blink, and fade the Chibitronics LED stickers in different patterns. It is programmable from virtually any device that has an up to date web browser using Microsoft's MakeCode Editor or using Arduino-style syntax with Chibitronic's LTC Editor.

Paired with conductive copper tape, Chibitronic LED stickers, and some other art materials, you can create a striking light sculpture that can be reprogrammed to display different patterns of light.

This project will walk you through how I created an electronic artwork. You can follow along to create a similar piece of art for your home, classroom, or makerspace, or you can simply learn how to use the Chibi Chip in a complex paper circuit and adapt the skills to your own work.

Materials

- Chibi Chip Microcontroller Board and Cable (chibitronics.com/shop/love-to-code-chibi-chip-cable)
- Computer, Chromebook, tablet, or smartphone to run Microsoft MakeCode (makecode.chibitronics.com)
- Conductive copper tape with conductive adhesive (a.co/iAGLoye)
- Cellophane tape
- Chibitronics LED red, yellow, and blue stickers (chibitronics.com/shop/red-yellow-blue-led-stickers-pack)
- Mat board
- Vellum paper (a.co/6EFF6Mr)
- Pencil, ruler, scissors
- Box cutter, cutting mat, cardboard
- Soldering iron and solder
- Safety glasses
- ¼ inch square dowel
- Hot glue gun and glue sticks

Optional

- PowerBoost 500C (adafruit.com/product/1944)
- Lithium Ion Polymer Battery (adafruit.com/product/328)
- 22 gauge hookup wire, red and black
- Slide switch (adafruit.com/product/805)
- Frame with glass, hanging hardware
- USB hub to power/charge your artwork

Build Your Artwork

Your artwork will have three layers. The bottom layer will be the circuit with LED stickers laid out in some pattern of your own design. The next layer will be a vellum paper overlay that will provide a nice diffused effect for the LED stickers. The top layer is a mat board that provides another design for the lights to shine through.

I designed a simple geometric pattern of cutouts to fit a mat I already owned. Eleven cutouts worked for my design, so I designed a circuit using eleven LED stickers. You can control up to eight LEDs from a single pin of the Chibi Chip, which will shine in any pattern you program.

The instructions that follow can be modified for your own design.

1. Cut a piece of mat board to the size and shape that you want your electronic artwork to be. I cut mine into a rectangle, approximately four inches by twenty four inches.
2. Draw two parallel lines on the back of the mat board using the ruler and pencil. Next, draw polygons along these lines, marking the negative space that you will remove with an 'X'.
3. Use your box cutter to remove the marked shapes from the mat board.
4. Cut a piece of vellum paper to the same dimensions as your mat board.
5. Cut a piece of cardboard to the same dimensions as the mat board.

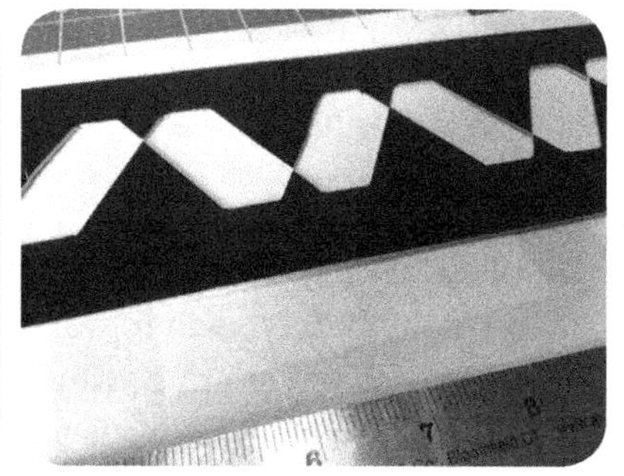

6. Place the Chibi Chip at one end of the cardboard. Use your pencil to start drawing your circuits. The circuits attach to the Chibi Chip "pins" which are the copper colored tabs at the bottom of the chip. This particular piece used a beta version of the Chibi Chip so the pins are in slightly different order than the final production version that is now being sold.

   For this piece, I wanted the Chibi Chip to light up three different colors of LEDs, each color group on a different circuit connected to the 0, 1, and 2 pins on the Chibi Chip. Each of the LEDs must also have a connection to the ground circuit on the far left of the Chibi Chip.

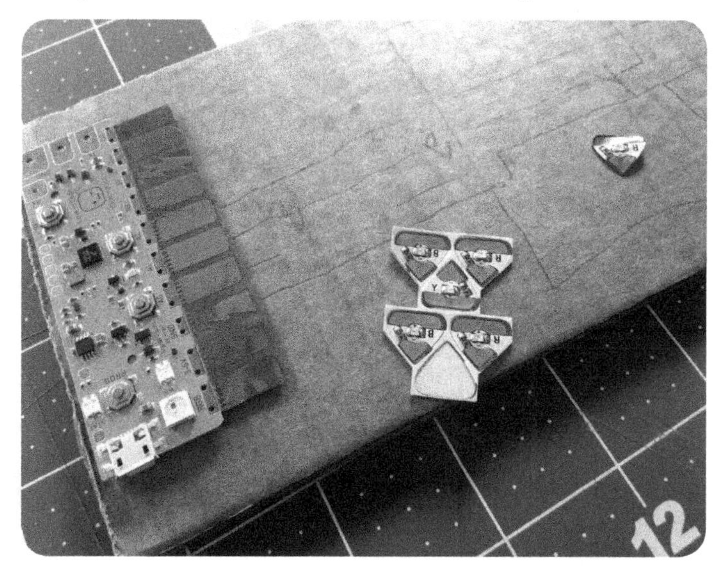

7. Here is a circuit diagram I drew to plan my circuits.

   The black line is the ground. The 0, 1, and 2 pins are represented by the color line that corresponds to the color LED in the artwork. Where circuits cross, the lower circuit is covered in cellophane tape to insulate it from the circuit above. The tape is represented by the gray hashmarks.

8. I split the ground circuit on my circuit diagram as I was laying it out, but later I noticed that I really didn't need to do that. There is no "one right way" to lay out a circuit, as long as each LED has access to power and ground, and the circuits don't cross. However, inevitably your circuits will cross. Note where these crossings occur.

9. Use a pencil to draw your circuitry on the cardboard, marking where you will affix the LED stickers and which color sticker (red, blue, or yellow) you will use.

10. Make sure your sticker placement lines up with the holes you cut in the mat board.

11. Read the Chibitronics copper tape tutorial web page and watch the embedded video for tips such as how to make the neat corner turns (chibitronics.com/copper-tape-tutorial). Following your pencil guides, affix the copper tape to the cardboard to build your circuits. Whenever a circuit needs to cross one of the other circuits, cover the copper tape with a piece of cellophane tape to insulate the circuit below.

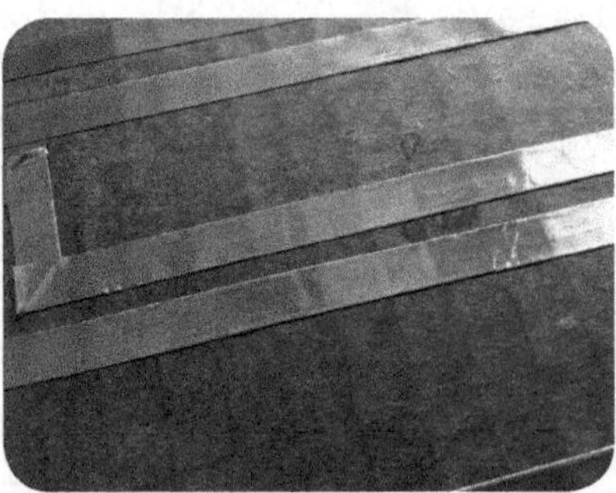

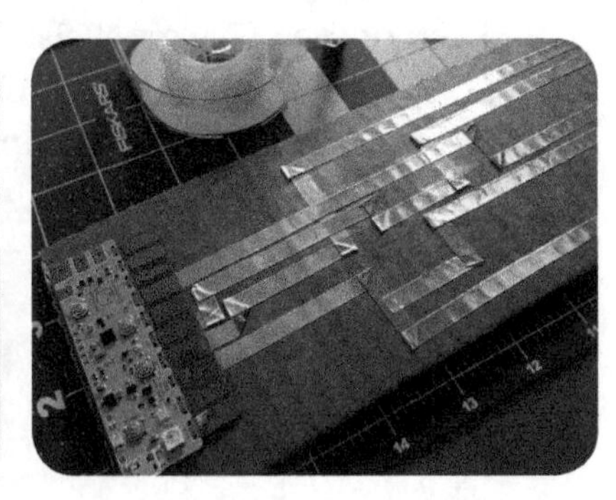

12. Adhere the LED stickers in the spots you marked on the cardboard. Make sure you mind the polarity of the circuit sticker, sticking the negative (-) side of the sticker to the ground circuit and the positive (+) pointy side of the sticker onto the appropriate circuit connected to one of the three pins.

13. Cover the pads on the circuit stickers with pieces of conductive copper tape.

14. Put on your safety glasses. Solder the copper tape junctions at each LED sticker. Additionally, solder the copper tape to the Chibi Chip pads. (Note: the LED stickers will work without soldering, but this project will be much more reliable in the long run if you solder them.)

Program Your Artwork

1. Connect the Chibi Chip to your computer with the supplied cable. The Chibi Chip is powered by USB and data is transferred via the audio cable. If you are using a smartphone or tablet, plug the USB part of the cable into a USB hub to power the Chibi Chip. If you are using a computer or Chromebook, plug the USB part of the cable into one of the USB ports. Connect the audio cable to the headphone jack of your computer or tablet.

2. Open Microsoft MakeCode for the Chibi Chip in a web browser (makecode.chibitronics.com) and create a new project. Assemble these blocks to test the circuitry you have constructed.

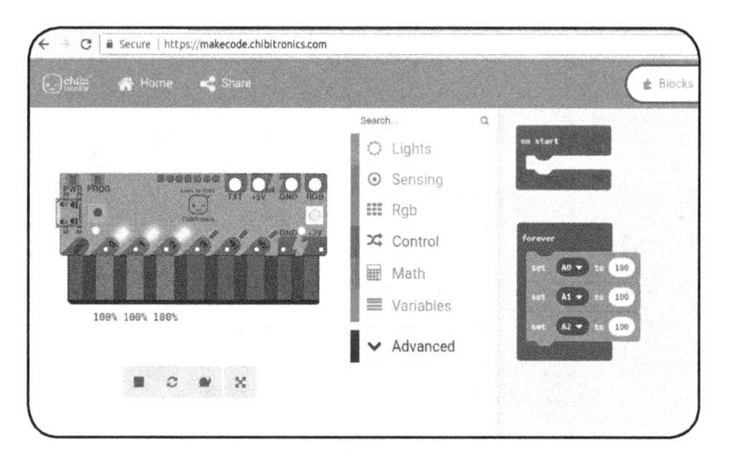

3. Click the Upload button to transfer the data to the Chibi Chip. Make sure the volume is turned up on the headphone jack. Press the button on the Chibi Chip to run the program. The LEDs should illuminate. If not, check your soldering.

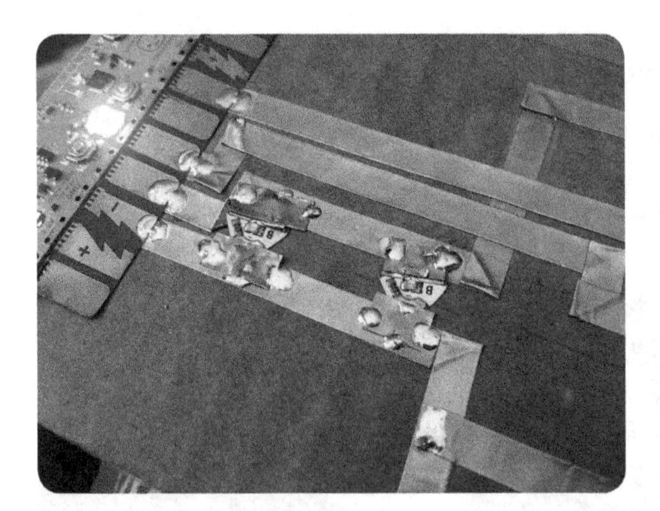
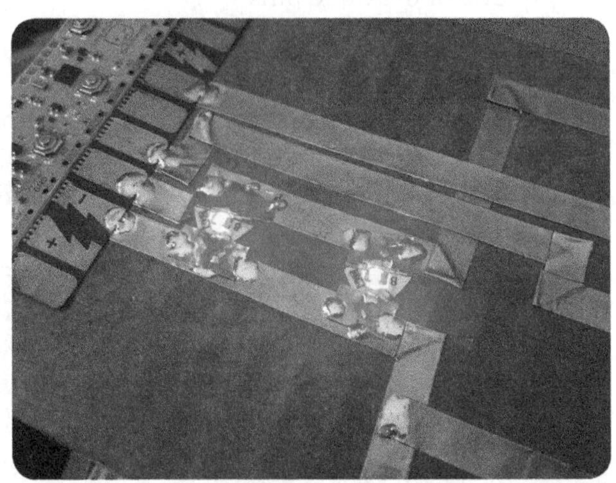

4. You can change the blocks to program a different pattern once you know all the LEDs light up. Upload the new program if you change the pattern. Test to make sure all your LEDs are working.

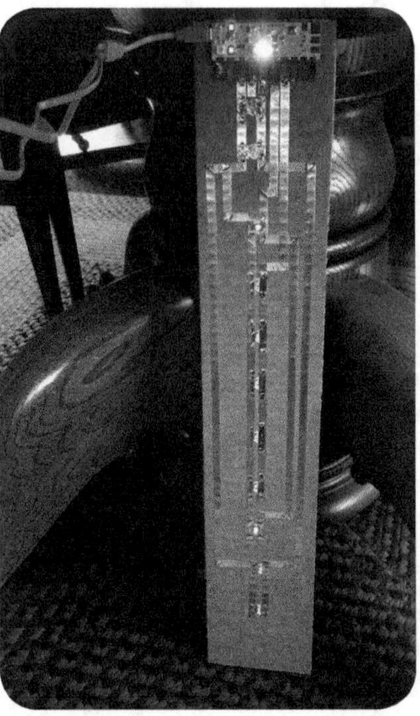
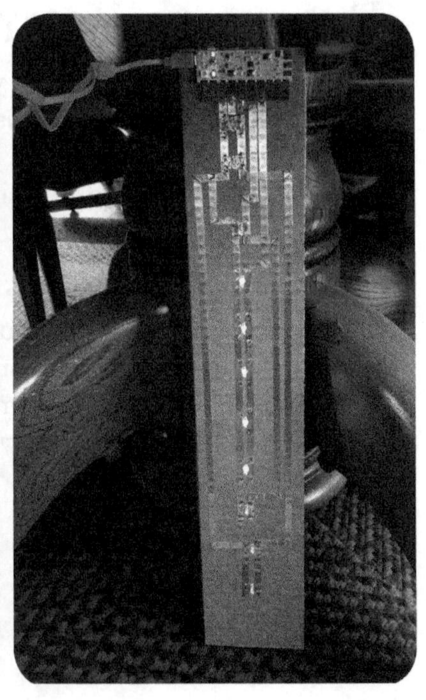

Final Touches

1. Cut two pieces of ¼ inch square dowel the length of your artwork. Hot glue them to the edges of the cardboard. Set the vellum and mat board on top of the dowels. The dowels set the LEDs back a little to make the light diffuse through the vellum more effectively.

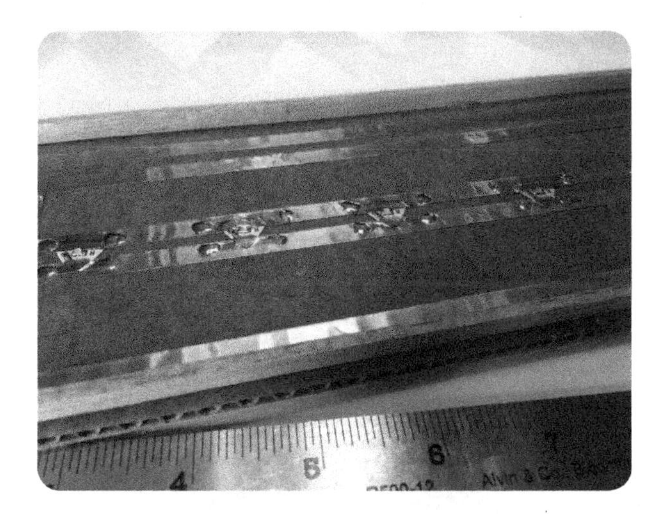

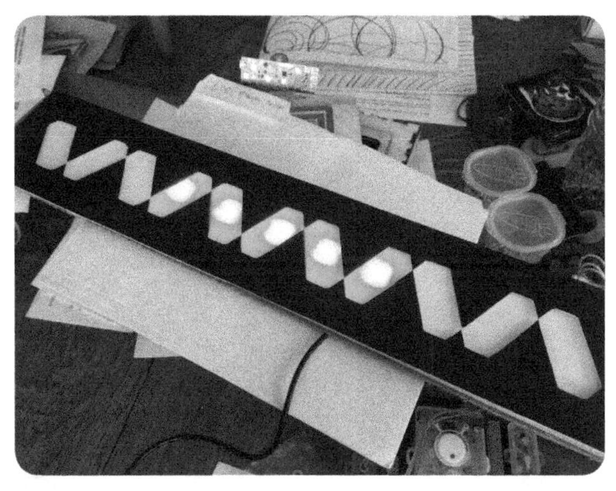

2. If you want this piece to be standalone and framed, now is the time to add the optional Powerboost 500C to your project. Solder a red wire to the +3V pad on at the bottom right of the Chibi Chip. Solder a black wire to the GND- pin on the upper right of the Chibi Chip.

3. Using the datasheet as a guide, solder the red wire to the positive input on the Powerboost 500C. Solder the black wire to the adjacent negative input. Solder red and black wires to the appropriate pins on the slide switch, and solder the other ends to the EN and GND pins on the Powerboost.

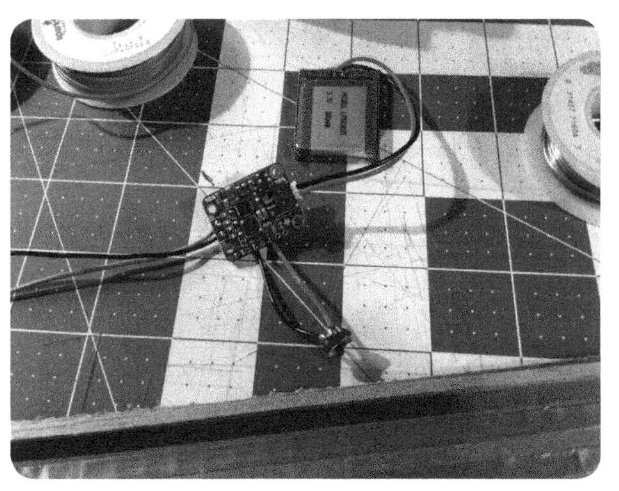

4. If you are having the artwork framed, cut out panels from the backing in which you can place the Powerboost and the battery. Plug in the battery to the Powerboost. Charge the battery through the Powerboost by connecting the Powerboost to a USB cable. The Powerboost will indicate with its LED when the battery is charged.
5. Use the framing hardware to hold the switch cable in place. Hot glue the switch to the back of the frame for easy access.

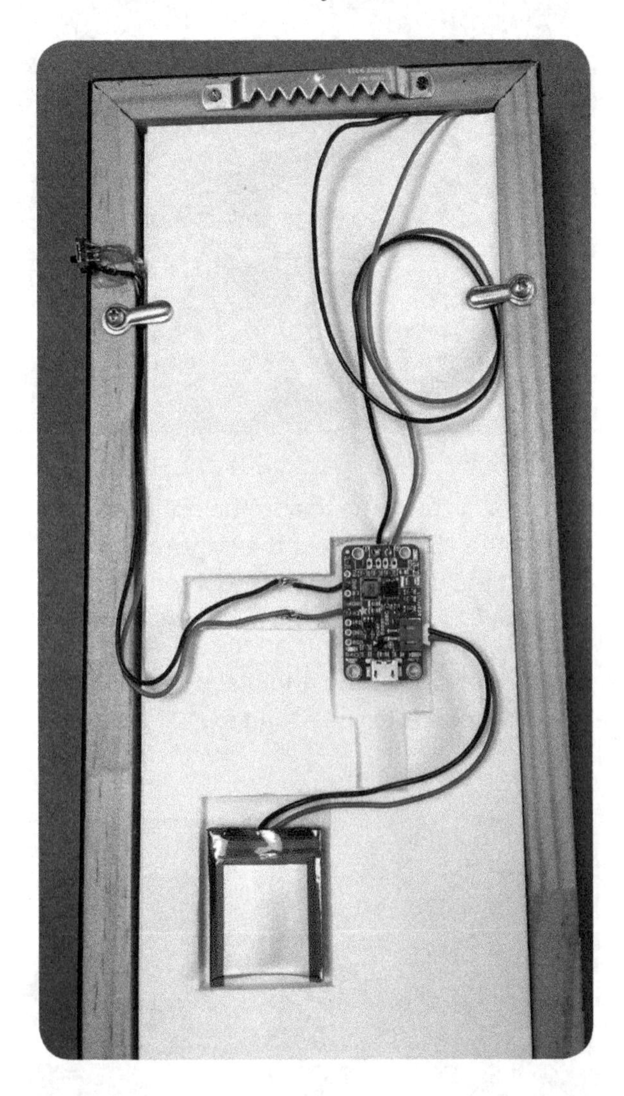

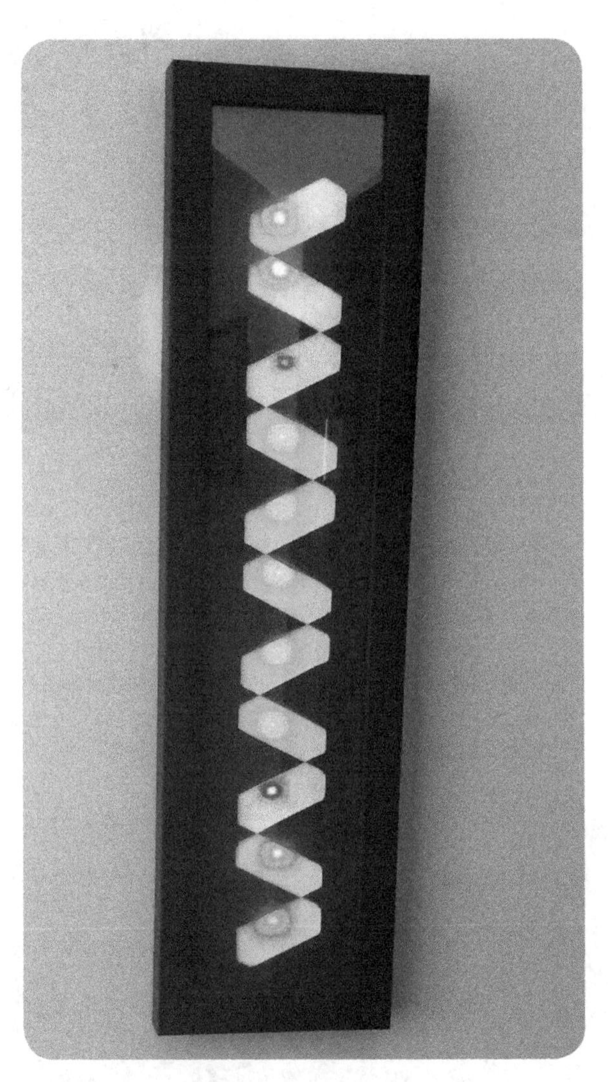

Whether you run your artwork during the day or at night, it will be a beautiful conversation piece. You can change the pattern by uploading a new program to the artwork, making it infinitely reconfigurable.

Passive Amplifier for Mobile Phones

Sometimes the most elegant solution to a problem is analog, not electronic. Mobile phones are convenient ways to access your music collection, but what if you need to share your jams beyond your ear buds? What if you are at the beach? Or all the outlets are in use at the maker space and besides, nobody has any speakers you can power and plug in to your phone? By crafting a passive amplifier from easy to cut and carve insulating foam, you can bring the noise anywhere you go with your phone and an amplifier that makes your phone's tiny speakers fill the room with sound!

Materials

- Insulating foam sheet (these come in 2 foot by 3 foot by 1 inch sheets at big box home improvement stores)
- Sharpie pen
- T-square
- Ruler
- Snap blade knife
- Cutting mat
- Cups in decreasing diameter sizes, or, optionally, the Windfire Designs Circle Tool (windfiredesigns.com/Tools/CircleTool/index.html)
- Rubber Bands

Optional

- Gorilla Glue
- Drop cloth
- Sand paper
- Spackling paste (preferably one that includes a primer)
- Nitrile gloves
- 2 inch spackle knife
- Spray paint
- Rubber bumpers for the underside of the amplifier

Build It

1. Working from a corner of the insulating foam sheet, use your smartphone as a guide to how large a square you should draw using the Sharpie and a T-Square. Size the square so your phone extends about halfway from the square if the bottom of the phone is at the middle of the square. For my iPhone 5, my square was about 4 inches wide and tall.
2. Using your ruler as a guide, carefully cut the foam using the snap blade. The advantage of using the snap blade is that you can gradually make the blade longer as you work your way through the relatively thick foam but still have control of the knife.
3. Cut two more squares the same size as the first. I cut an extra to practice cutting before I started the other three.
4. Use the cups (or the Circle Tool) to trace two circles, one larger than the other, on the insulation foam. Draw a wide outline of your phone on the third piece of foam, again positioning the phone about half way down the block of insulation foam. Use the snap blade to carefully cut out the circles, rounding the edges as shown if you like, or keeping them simple 90 degree cuts if you prefer. The third square should be carved out into a rectangular shape in which your phone fits loosely. You can stack the squares periodically to assess how they fit together. You want the phone to lean back slightly in the rectangle you carved so the speakers are tilted toward the "horn" you carved.
5. Connect the three carved squares using rubber bands as shown. Press play on your music, lower the phone into the passive amplifier, and notice how much louder it is!

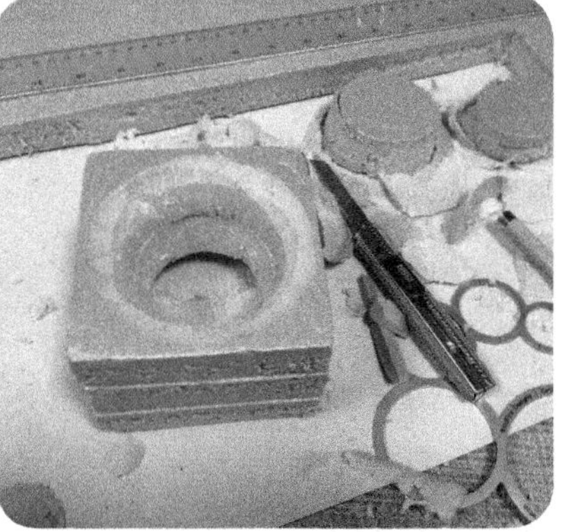

Painting and Finishing

Although the amplifier will work with only rubber bands holding it together, you can take this project further by gluing the layers together and applying different treatments to the surface. You can finish the foam with spackle and spray paint to make it appear as if it is made of something more durable than insulation foam. The finished effect, depending on your paint choice, makes your speaker look like it is made of an exotic material or metal. As always, consider that glues and paints are toxic substances and all safety precautions should be taken.

1. Apply a bead of Gorilla Glue to the edge of the squares to affix them to one another. You can use the rubber bands to hold the foam together while the glue cures. After the glue sets, use the sandpaper to finish off and smooth your carving and to rough up the surface a bit so the spackle will adhere to it better.
2. Put down a drop cloth. Put on your gloves. Prepare the spackle as recommended on the label. Apply an even layer of spackle to the body of the amplifier using the spackle knife. You might find it easier to apply spackle to the interior of the "horn" and the rectangle that holds the phone using your gloved fingers.
3. Wait for the spackle to dry and give it a light sanding. Apply a second coat, smoothing out any indentations or imperfections in the first layer. After it dries, give it a light sanding, too. Wipe off the amplifier with a dry, clean cloth.

4. Take your amplifier outside and put down a drop cloth. Following the directions on the can, prepare and evenly apply the spray paint to the model. For my amplifier I applied a coat of high quality metallic color spray paint, waited for it to dry, then applied a light coat of low quality black spray paint to produce an oxidized look.
5. After the paint is dry, turn over the amplifier and apply four rubber bumpers to the underside. This elevates the amplifier a little off the table and elevates the design, literally and figuratively.
6. Insert your phone into the amplifier and rock out!

Taking It Further

You could run a workshop on assembling these amplifiers for everyone's phones. It would be fun to create more whimsical shapes than a square, too.

My friend Reid Bingham, who inspired this project, builds his amplifiers out of wood using drills and saws. While the learning curve for these tools is higher, the results are more durable and as beautiful as a finished foam amplifier.

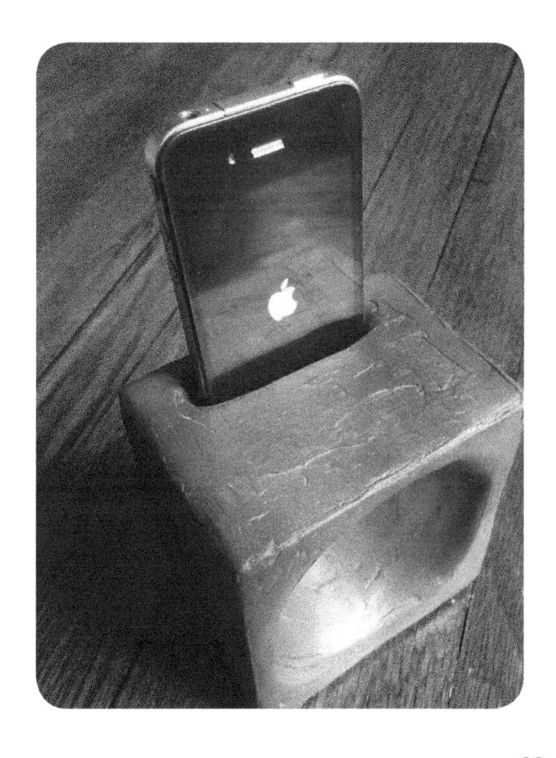

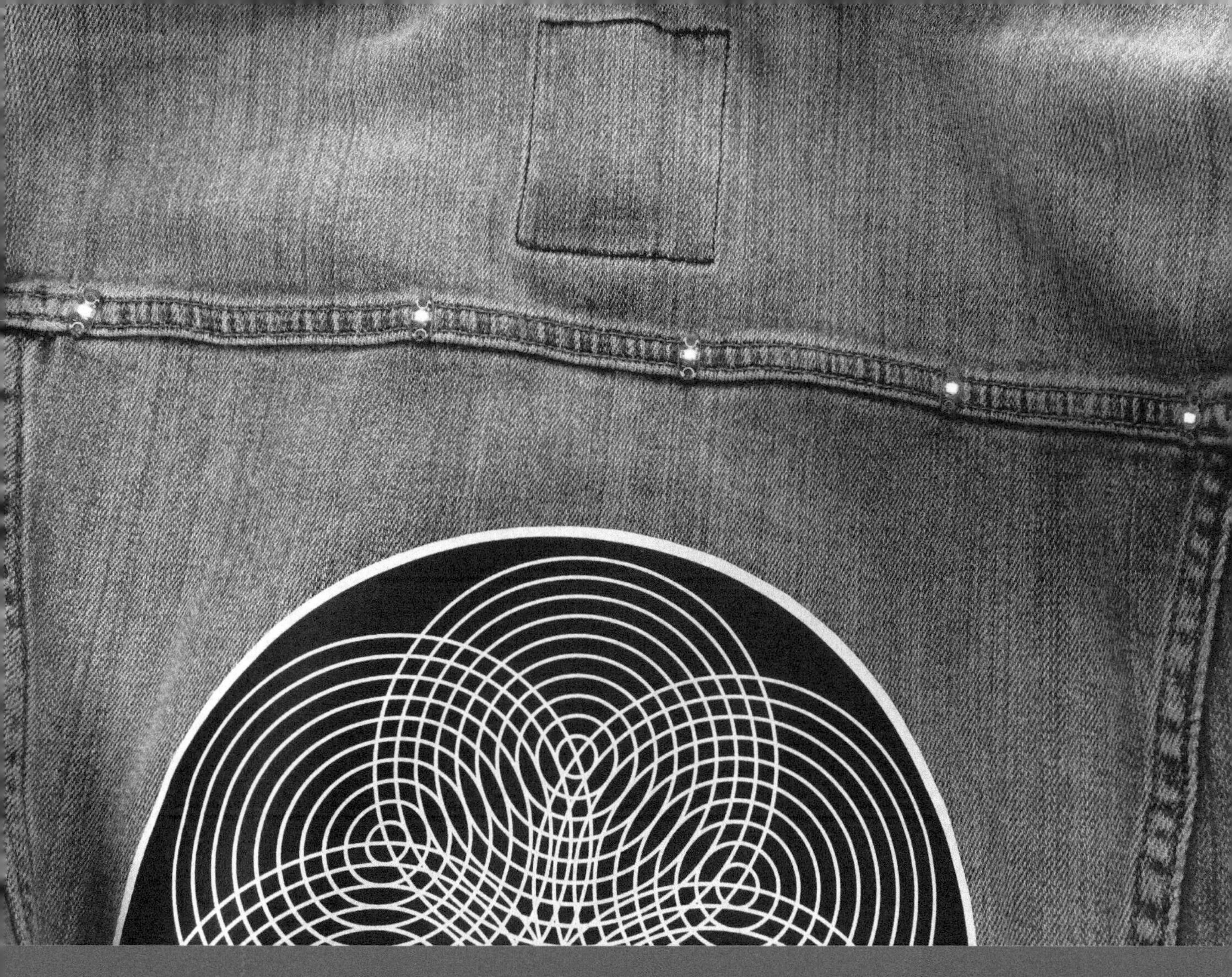

Light Up Jean Jacket

Walking the streets of Philadelphia after dark in search of toothpaste to replace the tube I forgot at home, two teenage girls standing on a corner watched me pass and make my way across the crosswalk. They shouted to me from across the street, "Mister, your jacket is lit!" and a big smile crept across my face. Indeed, with the help of the LilyTiny and LEDs, my jacket used soft circuits to flash five LEDs sewn across the back in a heartbeat pattern, lighting up my jacket and amazing everyone who sees it.

This project will help you customize a similar piece of clothing to my jean jacket. You do not need to use a jean jacket: I just thought it looked appropriately cyberpunk, both futuristic and vintage. This starter project will help you understand how the low cost LilyTiny microcontroller and LEDs can be added to just about any piece of clothing, powered with a simple coin cell battery.

I chose the LilyTiny as the microcontroller because it comes pre-programmed with four routines: blink, heartbeat, random fade, and breathing fade. This way you do not need to program the microcontroller and can simply illuminate your clothes with the circuit you sew. After this project you can move on to programming more sophisticated microcontrollers meant for wearables. This project is a low floor entry to wearable electronics that everyone can enjoy and accomplish.

The LilyTiny is a part of a family of Arduino boards called Lilypads. These boards were invented by Dr. Leah Buechley as a way to rethink cultural contexts for computing, and to broaden the appeal of microcontrollers to new audiences. Part of the proceeds of the sales of Lilypads go back to support further research in inclusive technology.

Materials

- Jean jacket or other item of clothing that you want to light up
- Conductive thread (sparkfun.com/products/10867)
- LilyTiny (sparkfun.com/products/10899)
- LEDs (sparkfun.com/products/14011) (these are green but they are available in red, blue, white, yellow, pink, and rainbow)
- Coin cell battery holder (2x2032 enclosed) (sparkfun.com/products/12618)
- 2 x 3V coin batteries (sparkfun.com/products/338)
- Sewing needle
- Scissors
- Seam ripper
- Hook-up wire (red - sparkfun.com/products/8023) (black - sparkfun.com/products/8022)
- Wire strippers
- Soldering iron and solder
- Safety glasses
- Electrician's tape
- Safety pin

Build It

Here is the circuit diagram of my light up jean jacket. This jacket uses a parallel circuit to illuminate the LEDs.

Front: LilyTiny & battery placement

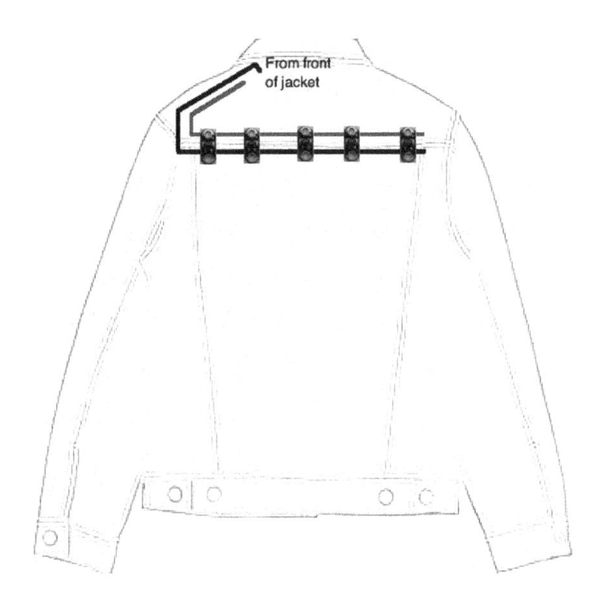

Back: LED placement

1. Start by deciding where you are going to place the LilyTiny in your jacket. The LilyTiny is literally quite tiny, and I placed mine inside the back of the jacket near the label.
2. Decide where you are going to place your LEDs. On this jacket I decided I was going to place them on the seam that runs across the back of the jacket at shoulder blade height.
3. If this is your first time using conductive thread to make a circuit, you may want to experiment with alligator clips in place of the wires and conductive thread before you start soldering and sewing your circuit into clothes; it will save you more complicated troubleshooting in the end.

4. Start by soldering wires that will connect to the battery. Put on your safety glasses. Solder a red wire to the + (positive) pad on the LilyTiny. Solder a black wire to the - (negative) pad. Using different colored wires helps you keep track of them, and it is standard practice to use red wire for the positive wire and black wire for the negative wire. These wires are sturdier than conductive thread and work well to connect to the battery.

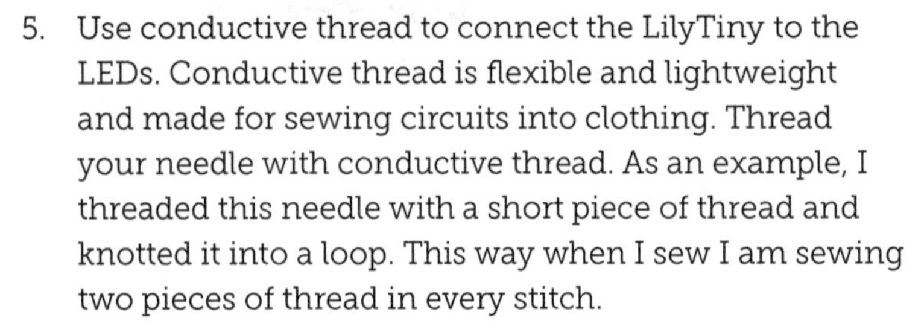

5. Use conductive thread to connect the LilyTiny to the LEDs. Conductive thread is flexible and lightweight and made for sewing circuits into clothing. Thread your needle with conductive thread. As an example, I threaded this needle with a short piece of thread and knotted it into a loop. This way when I sew I am sewing two pieces of thread in every stitch.

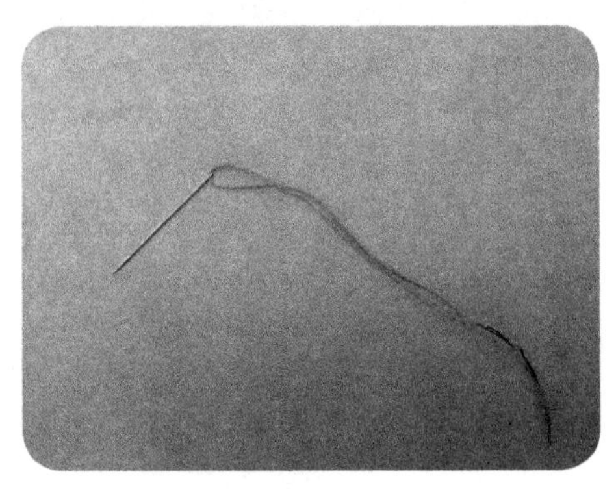

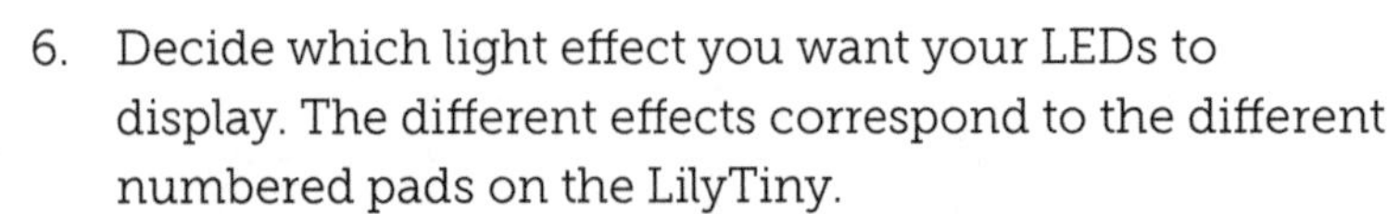

6. Decide which light effect you want your LEDs to display. The different effects correspond to the different numbered pads on the LilyTiny.
   - Effect pad 0 – Breathing fade
   - Effect pad 1 – Heartbeat fade
   - Effect pad 2 – Blink on and off
   - Effect pad 3 – Random fade
7. Sew through the hole in the pad that you selected. Run about four stitches through the hole to make sure it is secure and you have good conductivity.

8. Continue to stitch from the LilyTiny to sew the top part of your circuit. This is the red color on the circuit diagram and the positive leg of your circuit. Use small stitches that will not gap or sag. When you reach the place where you are going to attach the first LED, place the LED so the thread runs through the + (positive) hole in the LED. Run a few stitches through this hole before continuing on with the seam.

9. Continue sewing along the seam and placing the LEDs with the + (positive) pad towards the top of the jacket. Sew a straight line, connecting the + (positive) pad on each LED with the conductive thread. At this point you should only have the positive pads connected.

10. At the end of the seam tie a couple of knots in the thread and trim the excess.

11. Thread your needle again with conductive thread. This time you will sew from the - (negative) pad on the LilyTiny. Run a few stitches through the - pad hole so the thread is tightly connected to the LilyTiny. Sew towards the back of the jacket and connect each - (negative) pad of your LEDs. This thread run is represented by the black line in the circuit drawing. Be careful to stay far away from the thread connecting the positive pads or you will create a short circuit.

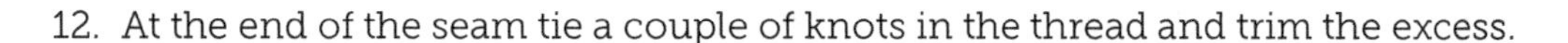

12. At the end of the seam tie a couple of knots in the thread and trim the excess.

There should be two parallel lines of conductive thread running across the back of the jacket. One thread runs from the effect pad of the LilyTiny to all the + pads of the LEDs. The other thread runs from the ground - pad of the LilyTiny to all the - pads of the LEDs. These two thread lines should not cross or touch or your circuit will not work.

Add Power

1. Next, you are going to place your battery pack. The 2x2032 battery holder is quite thin. I used a seam ripper to remove the seams at the top of the label inside the jacket. The battery holder slips right in.

2. Use the wire strippers to remove about 5mm of insulation from the battery holder wires and the wires you soldered to the LilyTiny. Put on your safety glasses. Connect the wires of the battery holder to the wires you soldered to the LilyTiny. Solder the red wire to the red wire and the black wire to the black wire.
3. Cut a short piece of electrician's tape and wrap the soldered joint. Repeat for the second wire.
4. Use a safety pin to secure the wires to the jacket so they do not get snagged as you put on the jacket. The safety pin contributes to the cyberpunk fashion, too.
5. Switch the battery pack to the on position.
6. The LEDs that you sewed into the soft circuit will illuminate!

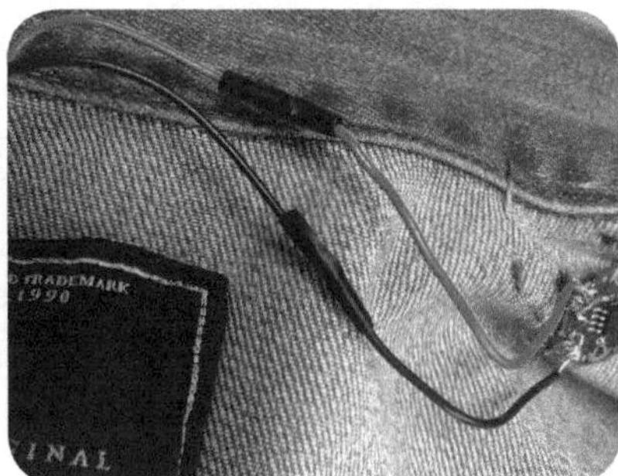

Taking It Further

As you see, sewing a parallel circuit and incorporating LEDs and lighting effects into your clothes is quite simple using the LilyTiny. I decided to make all the LEDs blink in the same pattern, but you could make a different choice by connecting each LED to a different effect pad. It's a little more complicated, since each LED would have to connect to a different effect pad with a different run of conductive thread, and all of these would have to be kept separate. Plan ahead and test your circuits before you sew them!

You can also graduate to using the Lilypad Arduino and program your own light patterns, or connect buttons, buzzers, sensors, or other electronic elements.

Now that you have make one wearable electronic jacket you can start augmenting all your clothes. Have fun and light up your world!

Beetle Blocks: 3D Turtle Geometry

Beetle Blocks (beetleblocks.com) is on online block programming environment that uses "visual code for 3D design." Based on Snap! (snap.berkeley.edu) and familiar to anyone who has programmed Scratch, models programmed in Beetle Blocks can be exported for 3D printing. Beetle Blocks is a fun way to explore the concept of bits to atoms: transforming digital objects created on a computer into physical, real world objects created through fabrication. Beetle Blocks is extremely flexible and powerful, but like Scratch there are many blocks and starting with Beetle Blocks, for some, is daunting.

These projects approach Beetle Blocks from a turtle geometry perspective. Like the Logo turtle draws its own path in two dimensions, the beetle draws its path in three dimensions.

I adapted the ideas for these projects from Natalie Freed's tutorial on how to create blocks in Snap! (nataliefreed.com/squares-spirals-zigzags). The three projects in this lesson build upon a shared shape, a seven sided heptagon. Of course, your Beetle Blocks projects could be a square, a octagon, or any other polygon you choose. Each projects builds on the prior, so work your way through this chapter in order and you will have a good understanding of how to program to design 3D printable designs.

The programming in these projects is both procedural and algorithmic. Creating a group of blocks teaches the beetle new moves, called a procedure, a cornerstone of the Logo programming language. Once the procedure is named, it can be used many times, or even in other procedures. Cynthia Solomon, one of the inventors of the Logo programming language explains that creating small tools that can be used in bigger tools means you can create wonderful things. An algorithm, as Brian Silverman, the inventor of TurtleArt explains is, "... to express a sequence of things that express itself over time." These projects use an algorithmic procedure that changes its position in space, its shape, and its size over time. Learning to program this way is applicable to other projects you might work on later in TurtleArt, Scratch, Snap!, or myriad other programming languages.

Because of these powerful programming capabilities, designing in Beetle Blocks is a bit different than other CAD programs, such as Tinkercad. The prints themselves are different, too; they seem almost organic in the shape and complexity of the generated STLs. I do not see Beetle Blocks as a replacement for my typical 3D design software, but rather a foray into a very rich set of materials for creating fantastic 3D designs, some of which might not be 3D printable but that allow me to further explore turtle geometry.

Getting Started

1. Load beetleblocks.com in your web browser. I recommend creating an account on the web site and logging in. Your work will be saved in the cloud and will be accessible from other computers and shareable.

2. Click the Run Beetle Blocks link at the top of the window.

3. From the "Project" menu (it looks like a document) select "Start tutorial." This short lesson leads you through the Beetle Blocks palettes and options.

4. Pay close attention to how manipulating the beetle's x, y, and z positions affect its movement.

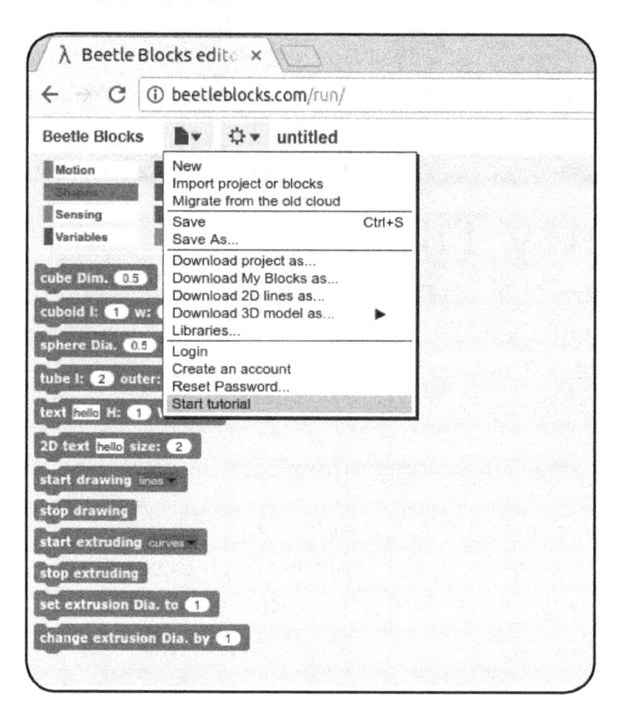

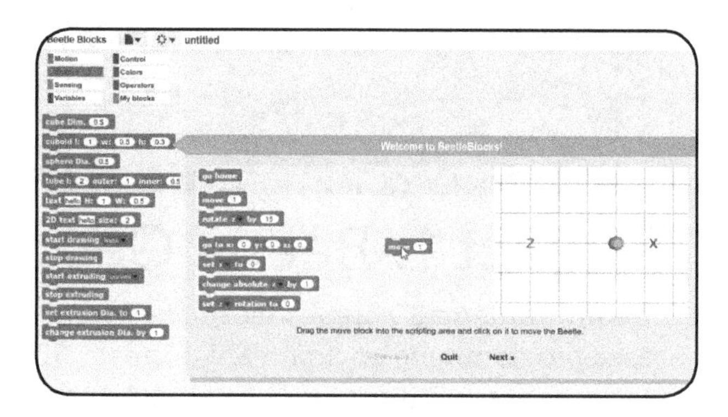

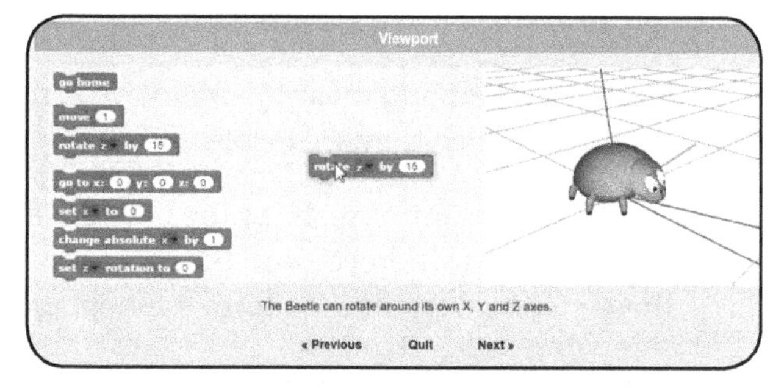

Just for fun...
Make a controller that allows you to "drive" the beetle in real time and in 3D space. All it takes is a Makey Makey and my Beetle Blocks Controller. It's a different way of exploring Beetle Blocks than a regular mouse. Try both and see what works for you! Beetle Blocks Controller plans:
labz.makeymakey.com/cwists/preview/307x

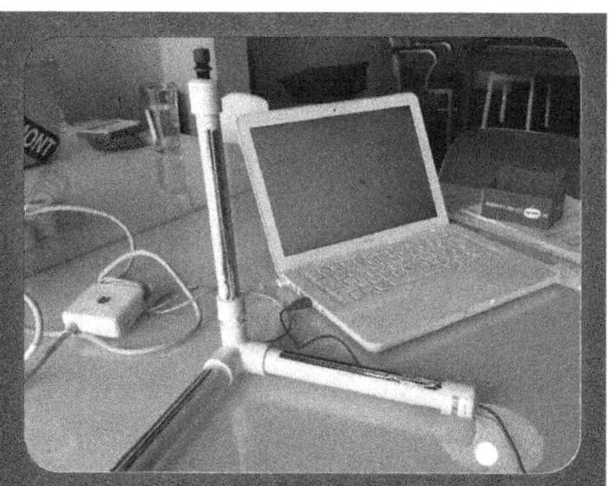

Project 1: Beetle Geometry Tile

Like my TurtleArt 3D printed tile in *The Invent to Learn Guide to Fun*, this project uses a block based programming environment to create 3D printable models. The models created in Beetle Blocks, as you will discover, have a much more delicate structure than models created with Tinkercad and are more suitable to display than to use as toys or tools.

1. Start by assembling these blocks in the scripting area. The colors of the blocks identify the palette in which they are located.

2. For illustrative purpose, I had the beetle draw with a pen (not extrude) the shape to show what happens when you run these blocks. The beetle moves five steps, turns, and repeats these steps seven times to form the heptagon.

3. Click on the My blocks palette in Beetle Blocks, at the upper left corner of the window.

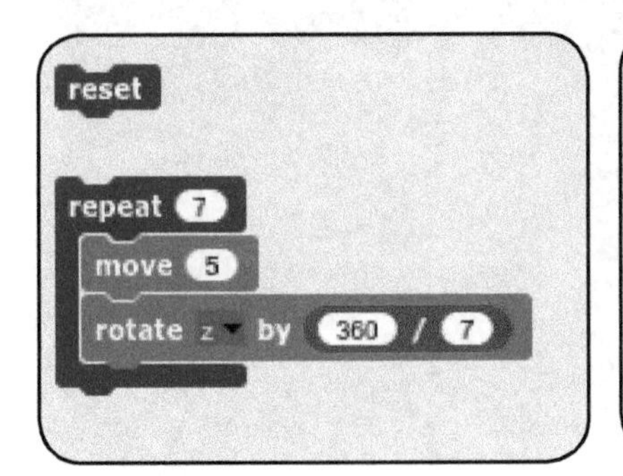

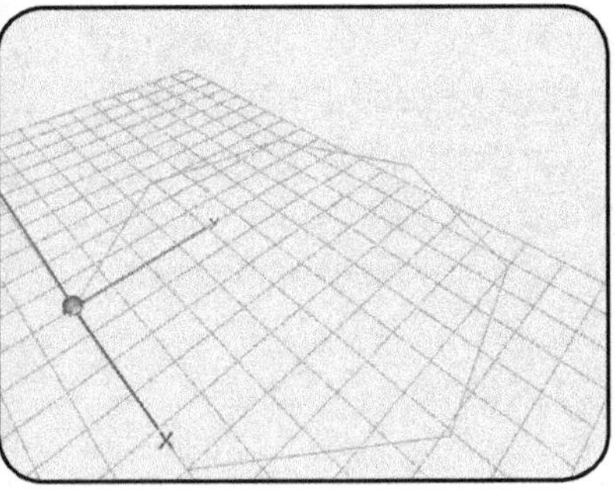

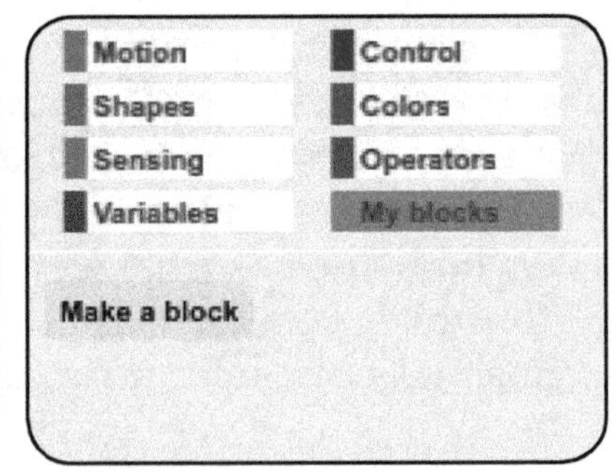

4. Click on the Make a block button to create a new block. This is akin to naming a procedure. Name the block heptagon.

5. Drag the stack of blocks you created in step one to the Block Editor window and connect them to the heptagon "hat" block.

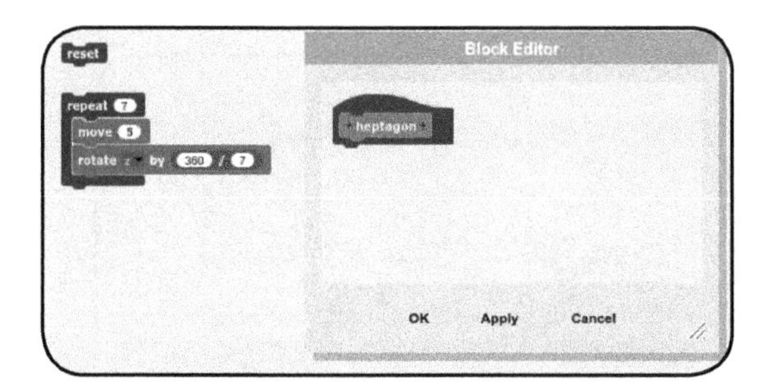

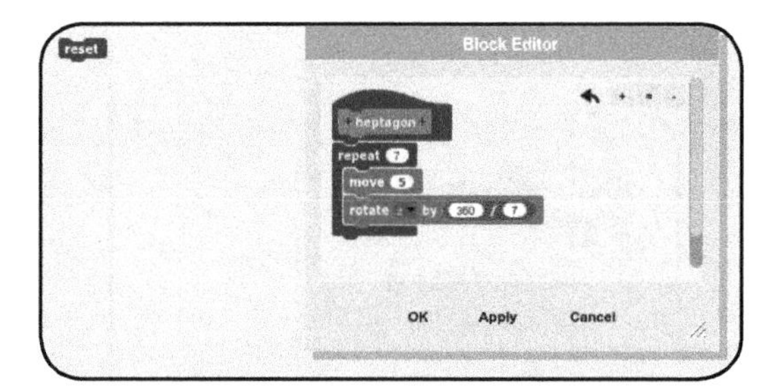

6. Click the Apply button, then the OK button. A new block, called heptagon, is created for you in the My blocks palette. Use this new block to create a heptagon anytime you want.

7. Click the green flag and you will see the beetle draw your design that uses the heptagon procedure.

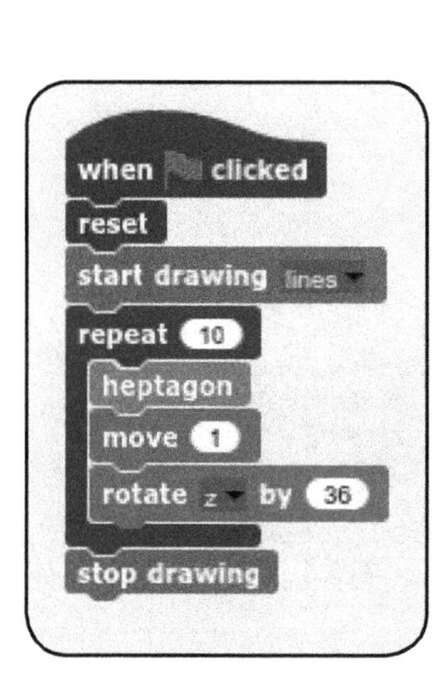

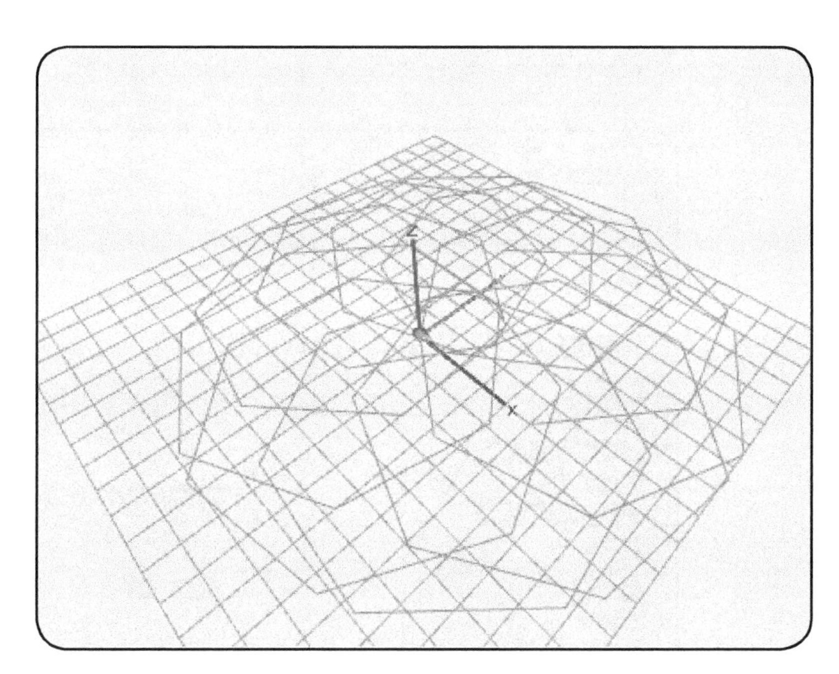

8. Now that we have tested our procedure, we will change it from drawing to extruding. This will create a physical object with thickness. We will use this procedure to create our 3D model.

9. Click the Green Flag. The beetle draws the design, this time extruding as it moves. The extrusion diameter is quite thin to preserve the details.

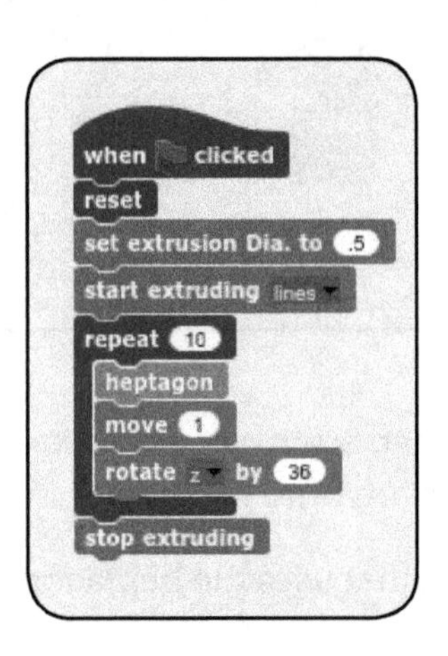

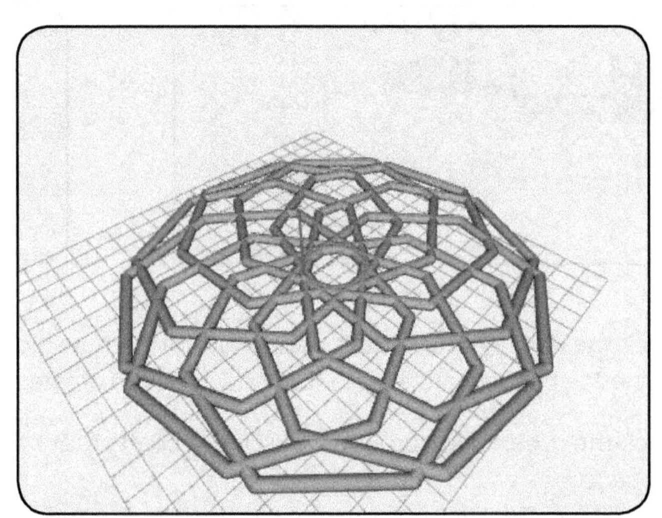

10. From the Project menu, select Download 3D model as... STL.

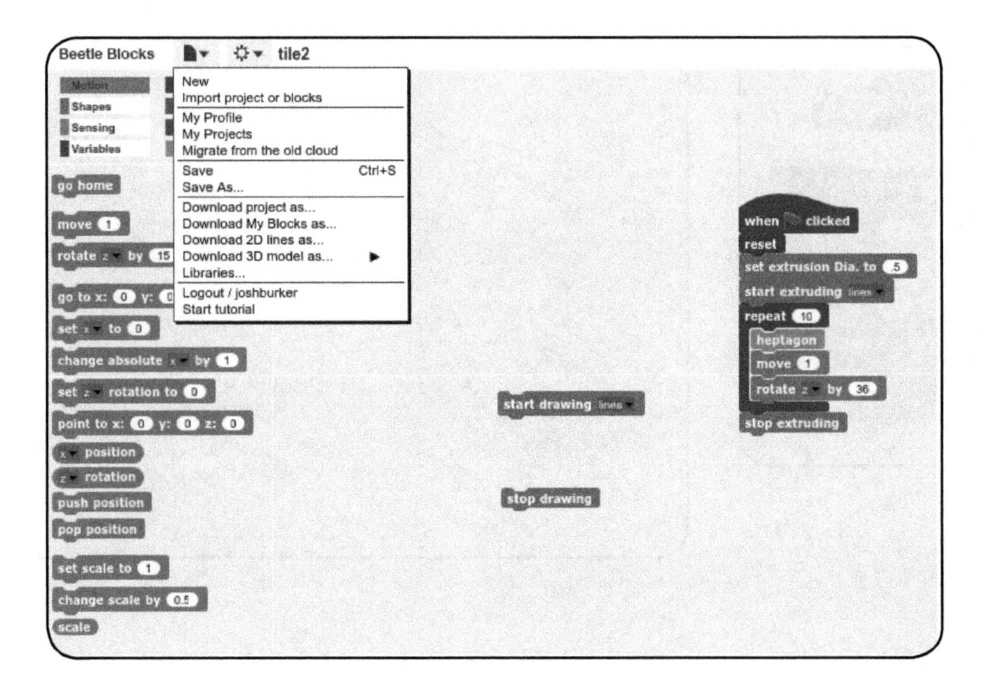

11. Load the STL into your 3D printer's slicing software. Slice the model (I used a 12% infill, .27m layer height).

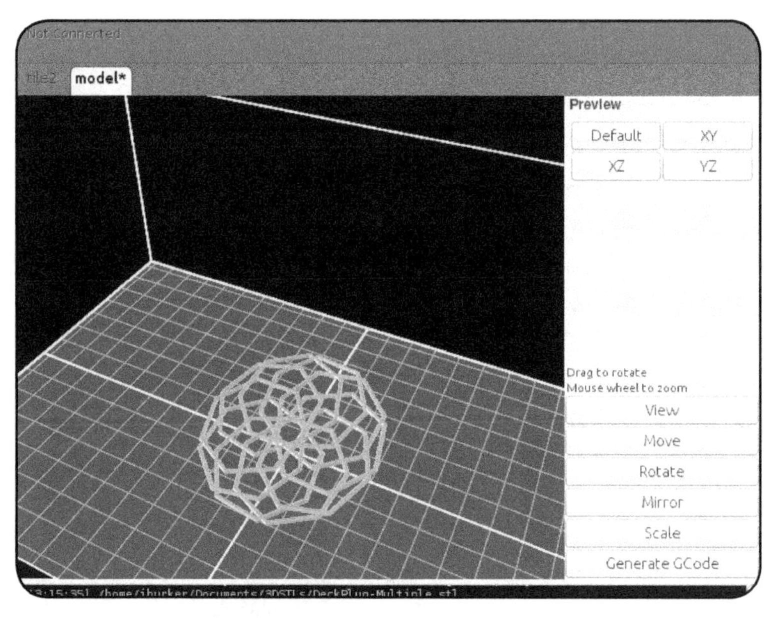

12. 3D print your Beetle Blocks model!

13. Once the model cools, remove it carefully from the printer. I find the models programmed in Beetle Blocks to be very organic in their shape and structure, with cylindrical extrusions and branching connections.

14. The models look beautiful framed with cardstock behind them.

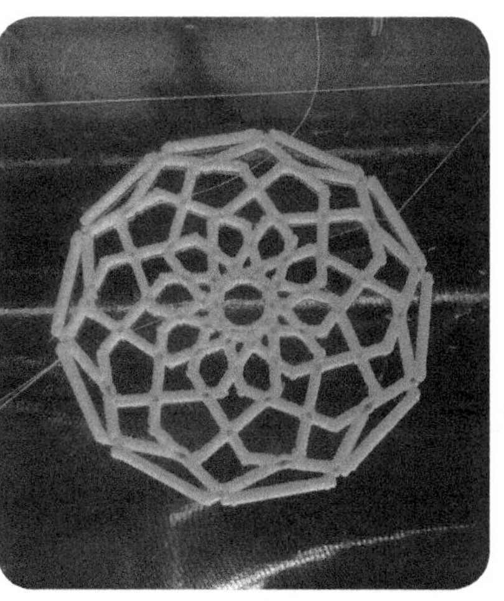

Project 2: Beetle Vase

This project literally builds on the heptagon design you programmed in Project 1 to create a 3D printed vertical vase. A one degree twist added to the vase makes the shape a little more interesting.

1. Start by creating a copy of your first project by going to the Project menu, selecting *Save As...*, and renaming the new file "vase."

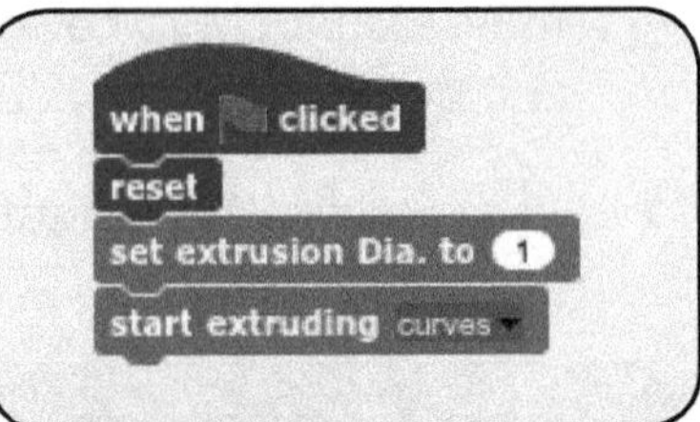

2. Take apart the blocks until these four are the only ones connected. Change the extrusion diameter to 1. Additionally, change the extrusion shape to curves.

3. Change the number of repetitions to 18.

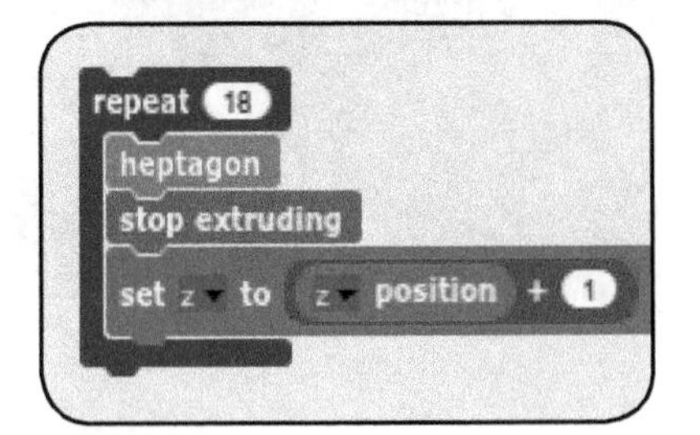

4. Add these blocks to the stack. These blocks cause the beetle to extrude the heptagon, then raise one level on the z axis and turn one degree. Doing this 18 times causes the vase shape to grow upwards.

5. Add this block to the bottom of the stack to make the design rotate 1 degree with each increase in height.

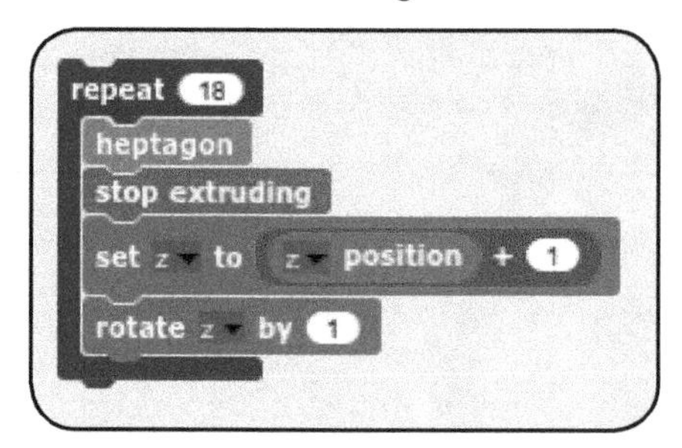

6. Add a block to start the beetle extruding curves again. I stop the beetle from extruding between vertical levels so there is not a big vertical line up the side of the vase. This block restarts the extrusion on the next level.

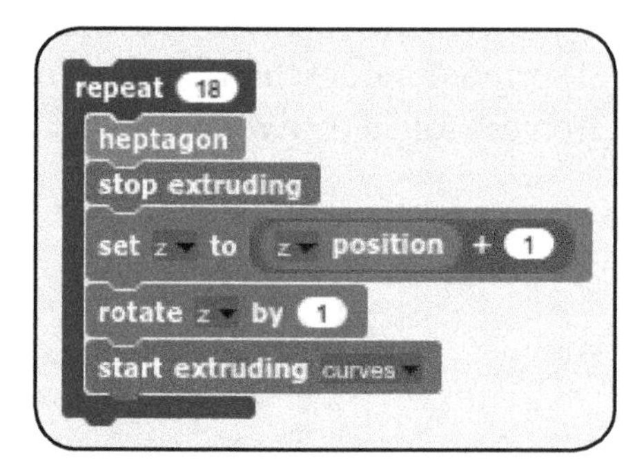

7. Connect the two stacks of blocks you programmed to create a single procedure.

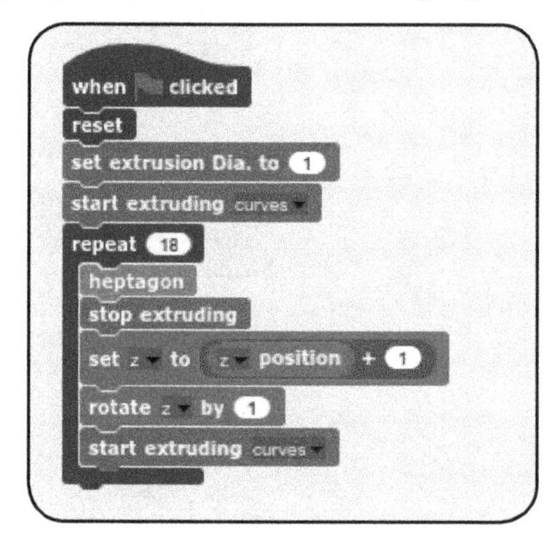

8. Click the Green Flag to run the procedure and create the model.

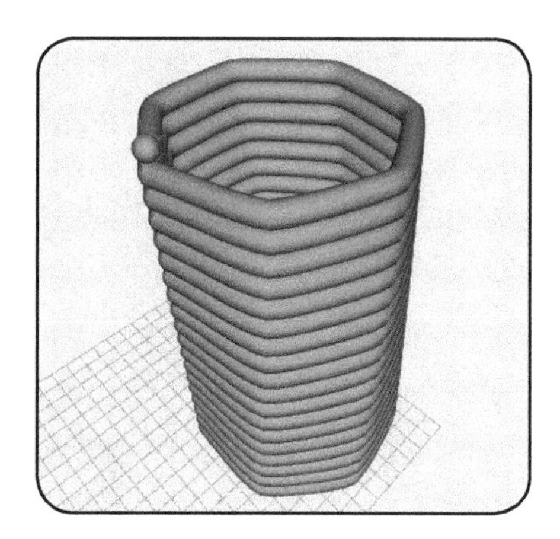

9. Save your work. Download the model as you did with the tile design and slice it in your 3D printer's software. 3D print the vase.

10. The completed model does not have a base. You can 3D print a base for it, hot glue the 3D printed model to a base made from different material, or leave it open.

11. A 3V coin battery and a color cycling LED inside the vase turns the model into a light vase. The relatively low 12% infill diffuses the light interestingly.

Project 3: 3D Printed Necklace

This project explores using variables in a remix of the heptagon block we created in the first project and used in projects 2 and now 3. We will create a 3D printed model than can be work as a necklace to display your programming and 3D printing prowess for all the world to see.

1. Start by creating a copy of your first project by going to the Project menu, selecting Save As..., and renaming the new file "necklace."

2. Go to the Variables palette in Beetle Blocks. Click the Make a variable button.

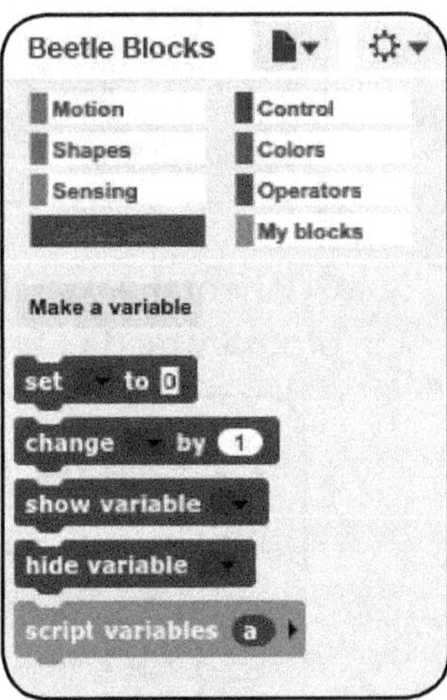

3. Name the variable "side."

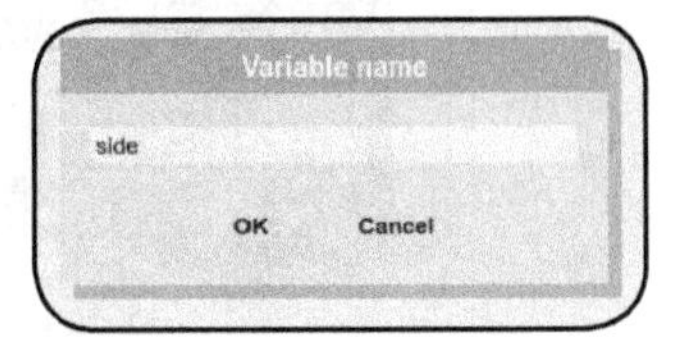

4. Go to the Operators palette.

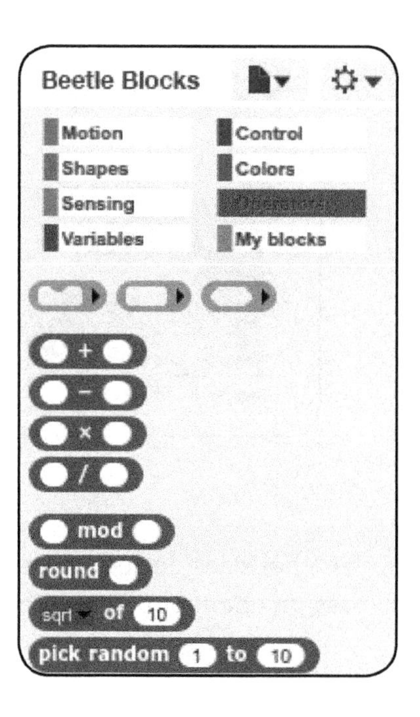

5. Use a combination of the pick random block and the set "side" variable block to create a block like the one shown.

6. Click on the My blocks palette. Right-click on the heptagon block and select edit.

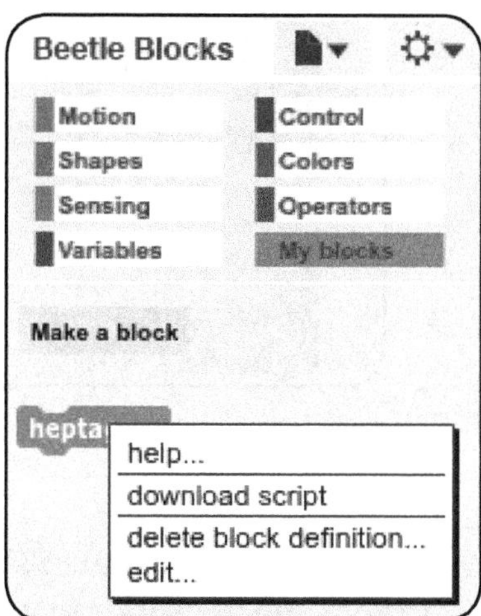

7. Change the heptagon block as shown in the Block Editor. We are randomizing the size of each heptagon the beetle draws. Click Apply and OK.

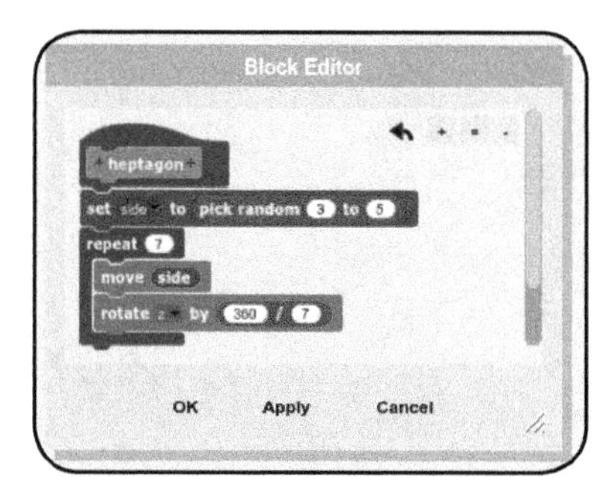

8. Construct these blocks to create extruded heptagons.

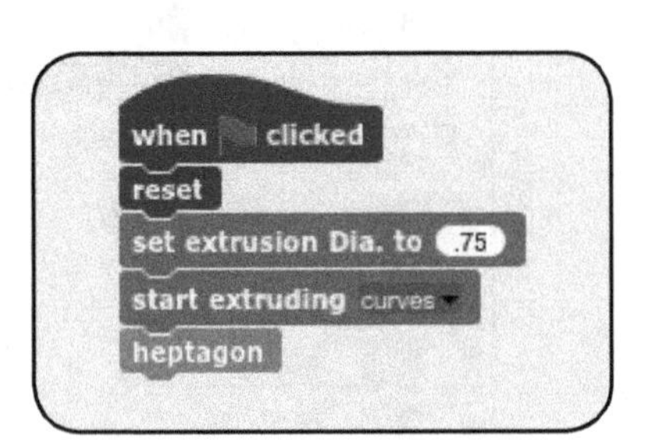

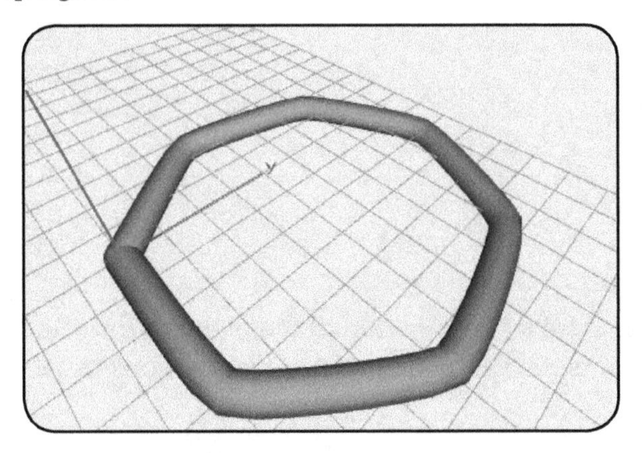

9. The necklace design procedure draws a series of heptagons rotating on the z axis. The design changes every time because the heptagon size is randomized.

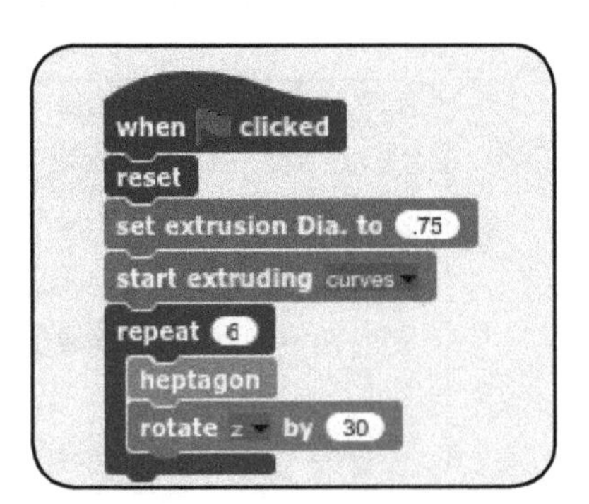

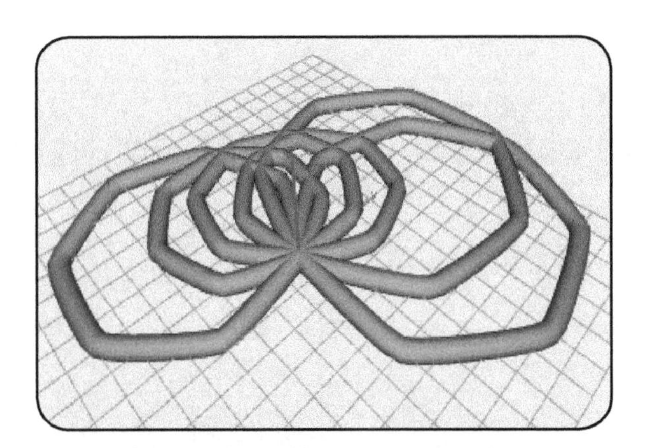

10. Save and download the STL. Size the model to your specifications in your slicing software. I wanted a fairly large pendant that I could hang from yarn.

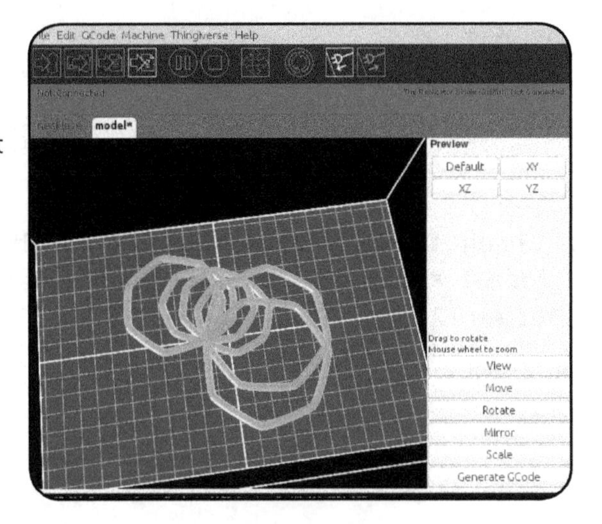

11. 3D print your model.
12. When the model cools you can remove it from the 3D printer and string it as a necklace.

Taking It Further

Now that you have a sense as to how Beetle Blocks works to program turtle geometry, start playing with other polygons. Remix these designs to have larger extrusion diameters, to be taller, or to have consistent sizes. There are many great projects shared on the Beetle Blocks site that explore other aspects of programming in Beetle Blocks and I encourage you to explore them, too.

Arduino Album Cover

Do you want to experiment with Arduino but are put off by the arcane Arduino programming language? Snap4Arduino may be your answer. Snap4Arduino combines the low floor approach of block-based programming with easy access to the Arduino hardware. Explore physical computing with the Arduino microcontroller and use your Scratch skills to rapidly prototype projects that illuminate LEDs, turn a servo motor, or read and write analog or digital data collected by the Arduino.

This project challenges you to paint an album cover, real or imagined, and to incorporate an LED. You will use Snap4Arduino to make the LED blink while the computer plays a music file and blinks a sprite on the computer screen as well.

Materials

- Computer running Snap4Arduino (snap4arduino.rocks)
- Arduino Uno
- 3D-printed Arduino Uno Mount, plus 2 x 5mm M3 machine screws to attach Uno to mount (optional, but highly recommended) (thingiverse.com/thing:33327)
- Assorted LEDs (5mm or 10mm LEDs work well for this project)
- 12" x 12" canvas (michaels.com/artists-loft-necessities-canvas-super-value-pack-12inx12in/10187423.html)
- Paintbrushes
- Acrylic paint in colors of your choice
- 2 x ½" wood screws
- Screwdriver
- 22 gauge solid core hookup wire
- Soldering iron and solder
- Safety glasses
- Wire snips/strippers
- USB cable to connect Arduino to computer
- Audacity (audacityteam.org) for music file format conversion

Paint Your Album Cover

1. Use your paintbrushes, acrylic paint, and canvas to create an album cover. Of course you can paint an album cover from a band you like, but it is more fun to be creative and paint your own creation.

2. Clean your brushes and wait for the paint to dry.

Add Electronics

1. Once the paint is dry, choose where on your painting you want an LED. Poke the legs of the LED through the canvas so the LED sits flat against the canvas. I painted a picture of my album cover (the front of a Mac Classic) and will use this project to illustrate the rest of the process.

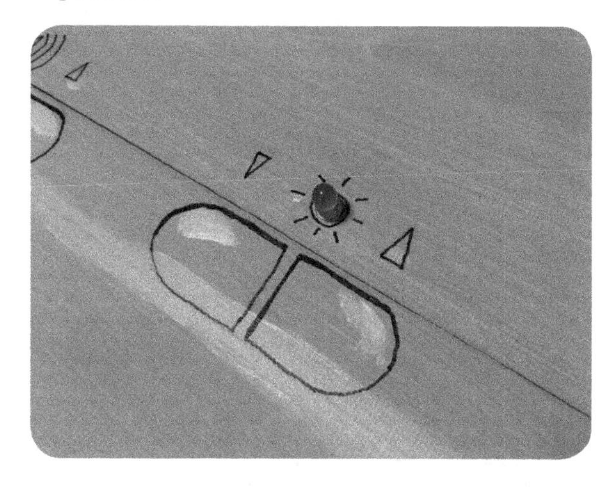

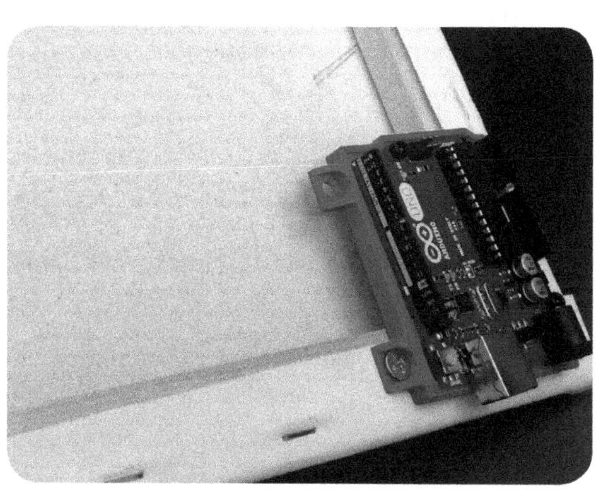

2. Mount your Arduino in the 3D printed Uno mount. Use two woodscrews to attach the mount to the back of the canvas.

3. Use the hot glue gun to secure the LED to the canvas from the back, leaving enough of the LED leads exposed that you can solder a red wire to the positive (longer) lead and a black wire to the shorter negative lead. Leave enough wire to comfortably reach the Arduino. Strip about one centimeter of insulation from the other end of both wires.

4. Plug the red wire into digital pin 9 on the Arduino. Plug the black wire into one of the GND connections on the Arduino.

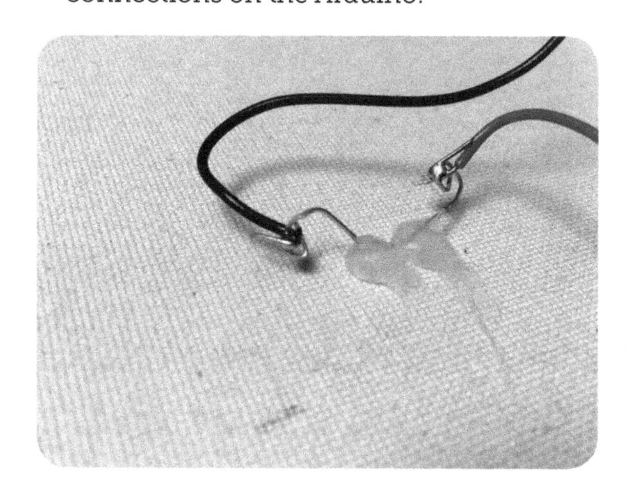

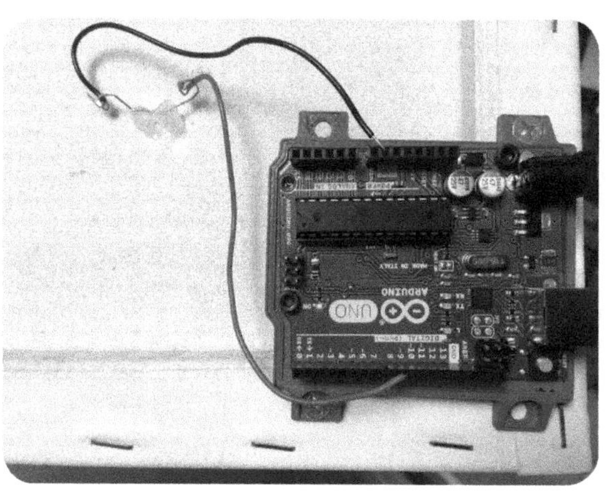

Program the Arduino Using Snap4Arduino

1. Connect the Arduino to your computer with a USB cable. Open your web browser and navigate to snap4arduino.rocks.

2. Download Snap4Arduino for your computer. The great thing about this program is it works on nearly every platform, from Chromebooks to Mac and Windows as well as Linux! The Snap! programming language is a variation of Scratch, so it may be a familiar programming environment.

3. After installing Snap4Arduino, open the application. Click on the Arduino blocks button. Click the Connect Arduino button and select a port (I chose /dev/tts.usbmodem1D11).

4. Snap4Arduino will connect to the Arduino.

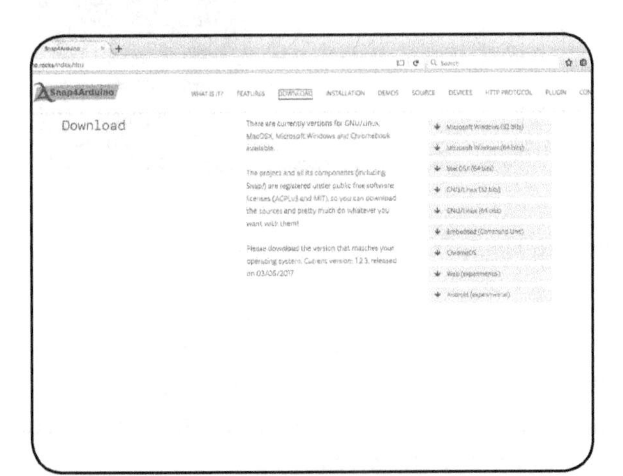

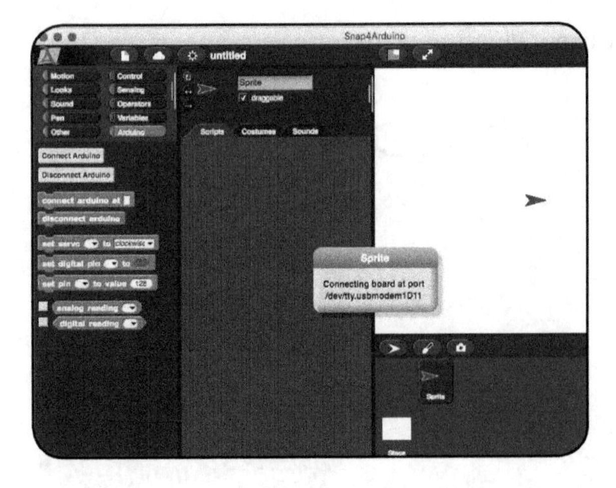

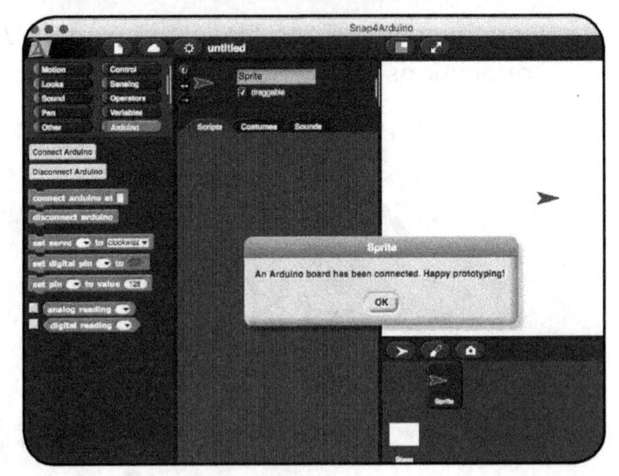

5. First, let's make sure the Arduino communicates properly with Snap4Arduino. Assemble these blocks. Type the number '9' in the pin variable slot, then you can click to slide the slider to 'on.' This command tells the Arduino to turn power on to what is attached to pin 9.

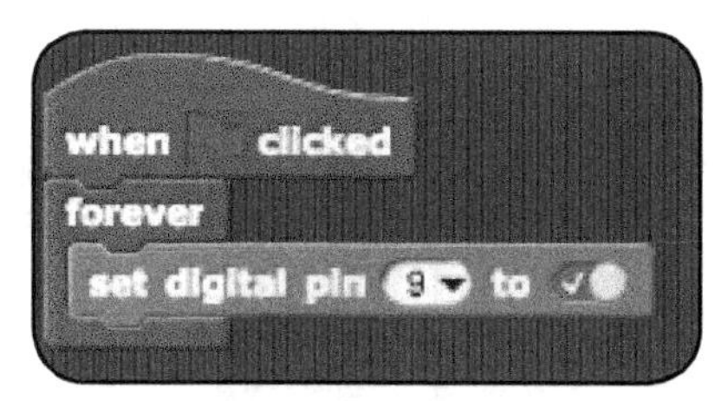

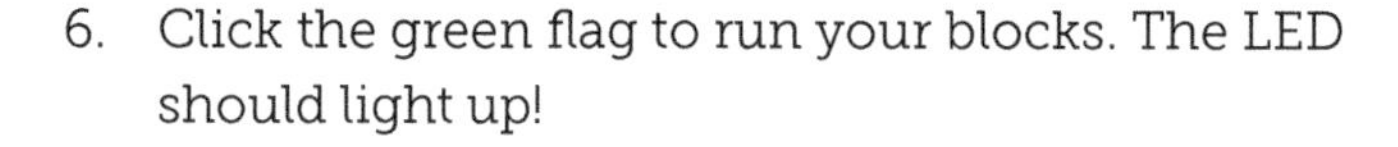

6. Click the green flag to run your blocks. The LED should light up!

7. Now, let's turn our focus to improving the Snap project. First, take a photo of your album cover and import it to the Stage as one of the backgrounds. You can simply drag and drop the image into the Backgrounds tab to add it to the project.

8. Next, let's make a Sprite that looks like the LED that is in your real world painting. Click on the Sprite, then click on the Paint a new costume button.

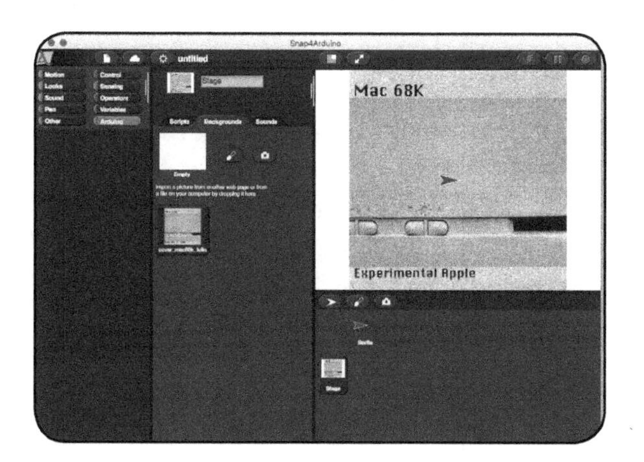

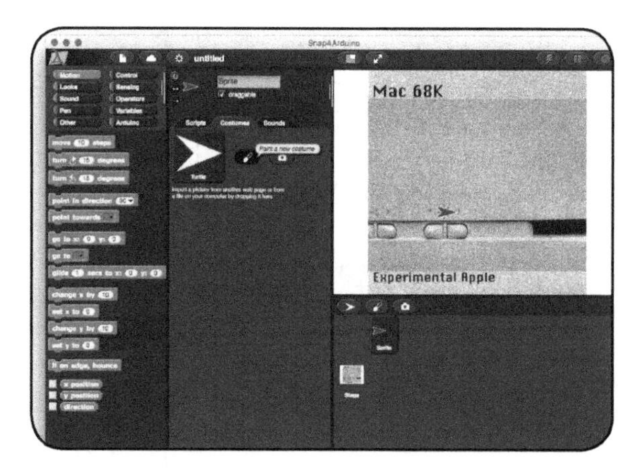

9. Use the paint editor to make an LED-looking dot.

10. Move the Sprite to the appropriate place on the Stage so it looks like your real world album cover and LED.

11. Now would be a great time to save your work if you have not already. You work can be saved locally on your hard drive, or in the cloud if you create a Snap4Arduino account.

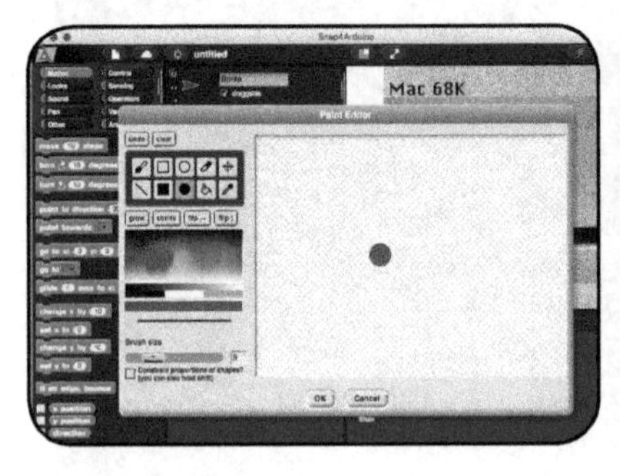

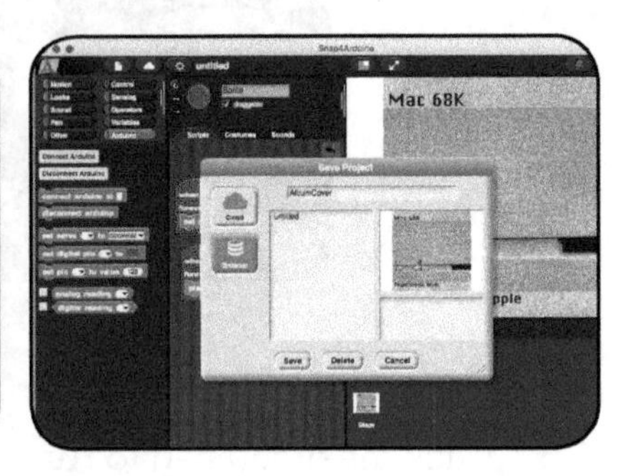

Add Music

1. Next, we are going to add a song to the project. Most songs are in MP3 format, but at the time of printing this book there is a known bug that prevents MP3s from playing in Snap4Arduino (github.com/bromagosa/Snap4Arduino/issues/30) so we are going to use Audacity to convert your song to Ogg Vorbis format. Open Audacity, then open the music file.

2. From the File menu, select Export Audio.

3. From the Formats menu, select Ogg Vorbis Files. Name your document and click Save.

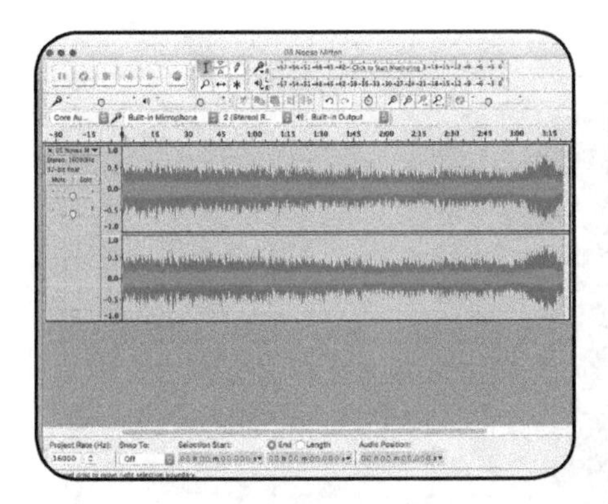

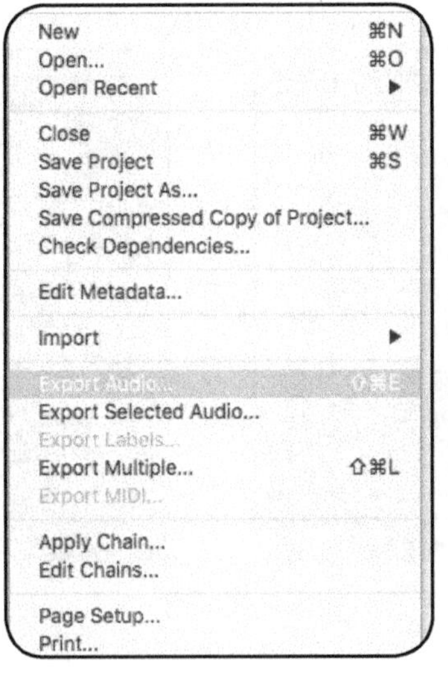

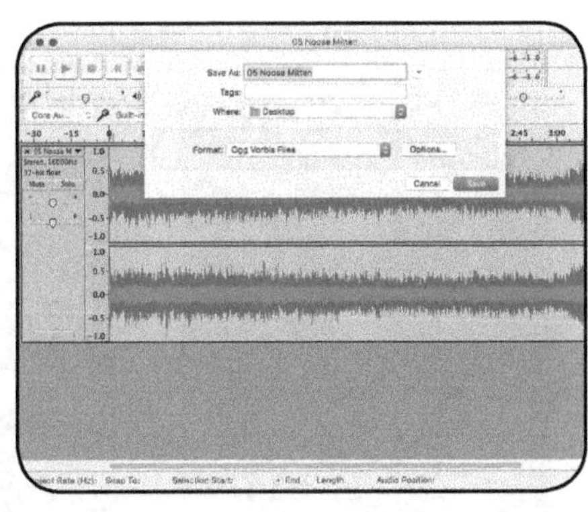

4. Drag the Ogg Vorbis file you created into the Sounds tab in Snap4Arduino to add the sound to your project.
5. Connect the Flag start Control block to the Sound block to play the sound.

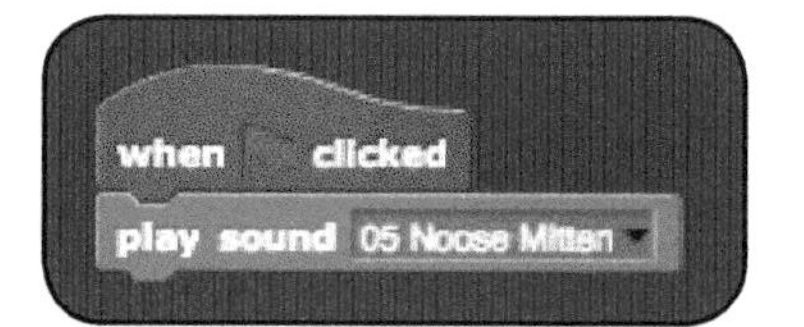

6. I modified the LED script to make the LED turn on and off, and I changed the Sprite script so it shows and hides, making it blink, too. Here are the completed scripts.

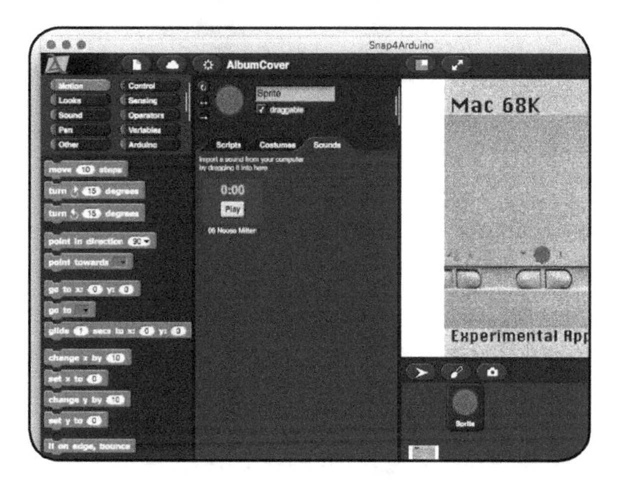

7. Click the Green Flag to start the project. All three stacks will start at the same time. The LED will blink, the Sprite will "blink," and the song will play.

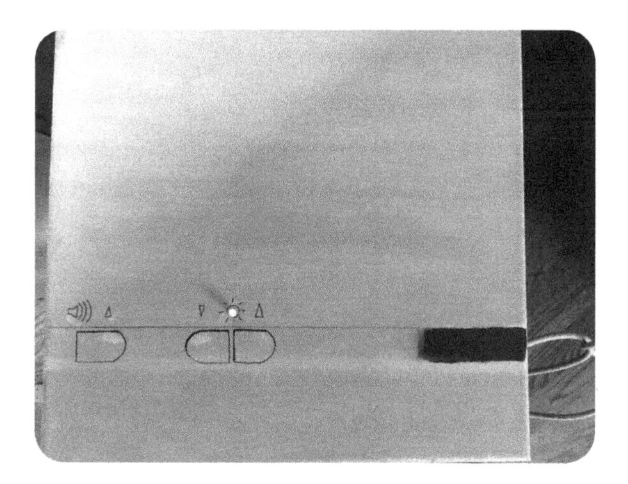

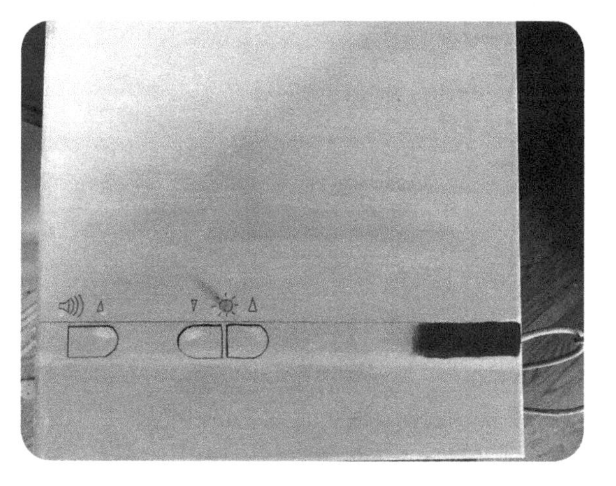

Troubleshooting

If Snap4Arduino loses connection with the Arduino, you can try switching which port it is looking for the Arduino. I was able to use either port to successfully communicate with the Arduino.

Taking It Further

Try adding additional LEDs. This student had multiple LEDs representing different cities in his painting. Try adding a servo or motor to make part of your album cover move. Add a light sensor so the album only plays at night.

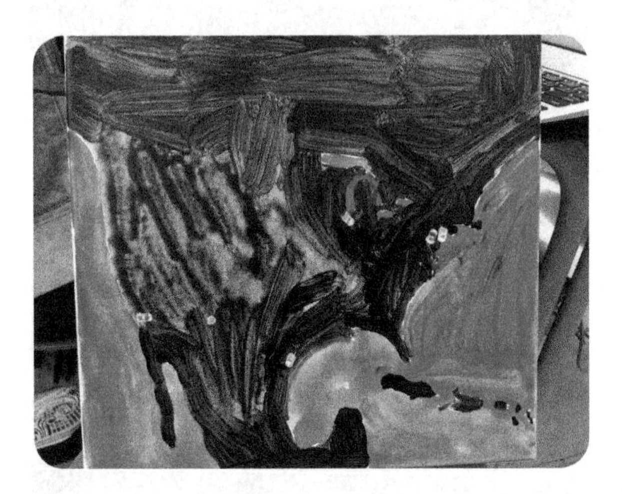

Snap4Arduino is a great low-floor introduction to the Arduino. You can do much of what you would do in the text-based programming language interface (called the IDE) but with block-based programming. Once you have a sense of how to integrate an Arduino into other projects, you can keep using Snap4Arduino or move to the IDE.

Turtlestitch Embroidery

Turtlestitch is an online programming environment similar to Scratch that creates designs that you can load onto an embroidery machine. Your Logo-inspired designs can be transformed from bits to atoms through the "magic" of this software. Although not every design can be embroidered because of the limitations of the machine, the material you are embroidering, or the thread, there is a wide selection of designs that can be sewn. This chapter will lead you through the basics of creating a design that you can embroider.

Materials

- Computer running Turtlestitch in a web browser (turtlestitch.org)
- USB drive
- Embroidery machine with the ability to import user generated files. I used the Brother PE-770 (a.co/cHCEa8P)
- Embroidery hoop set (a.co/gHVHTBm)
- Pre-wound bobbins (a.co/gWNp7MH)
- Embroidery thread (a.co/9olGT4f)
- Embroidery needles (a.co/eqFrnKe)
- Stabilizer backing sheets (a.co/3T3qpme)
- Dritz Fray Check (a.co/dN4w9Fw)
- Fabric or clothing that you want to embroider

Getting Started with Turtlestitch

1. Open Turtlestitch (turtlestitch.org) in your web browser. Create an account on the site; this allows you to save and share your projects from the site and allows you access to your designs from any computer.

2. The Turtlestitch team produced a set of "getting started" cards that are very helpful in orienting yourself to the limitations and design principles of programming for embroidery. I highly recommend you read through the cards; they are very well produced and will help you avoid later frustrations when it comes time to fabricate your design.

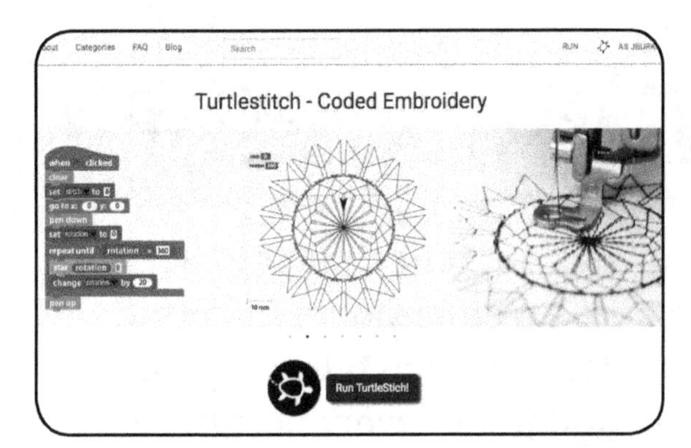

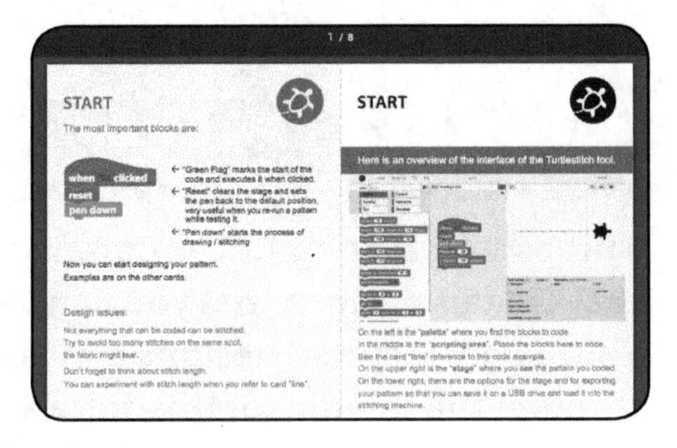

3. Run Turtlestitch by clicking the button in the browser window.

Create Your Design

Turtlestitch embroidery works best if the turtle moves in "chunks" of ten steps. Program your turtle's movements using the special movement block that breaks long distances into groups of ten.

It is important to note the size of your design in relation to the slightly darker part of the grid where the turtle draws. The embroidery machine has physical limitations to how large of a design it can embroider because of the size of the hoops that hold the fabric and backing. Plan your design accordingly and try to keep it within this grid, which translates to about 10 centimeters square.

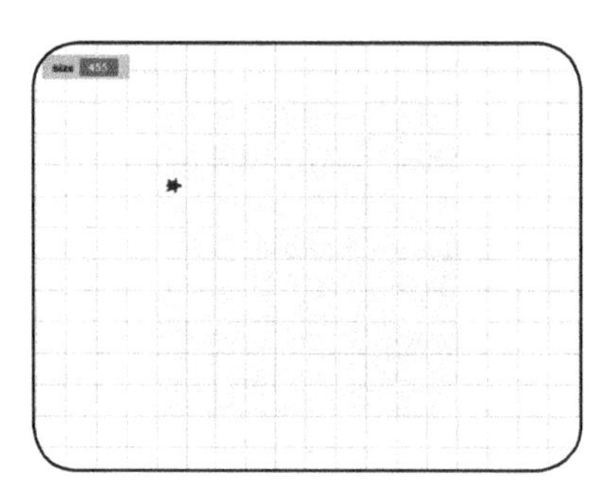

Run and debug your design. Try to keep the turtle from drawing too many times in a particular spot (the machine can end up ripping the fabric).

Embroider Your Design

When are happy with your design, download it to a USB drive that you can insert in the embroidery machine. The Brother PE-770 that I use accepts DST files as patterns.

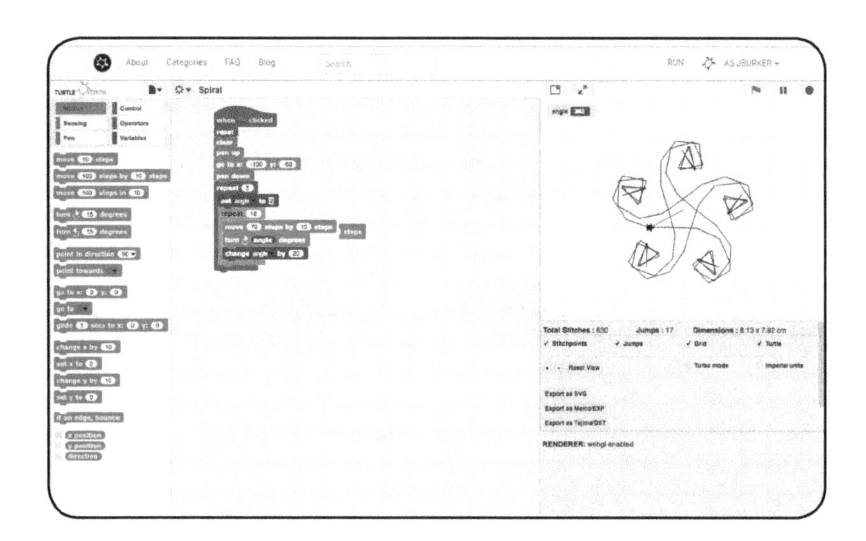

1. Click the *Export as Tajima/DST* button in the Turtlestitch window, below your design.
2. Insert the USB drive in your embroidery machine. Follow the manufacturer's directions to load the file from USB into the embroidery machine's memory.

3. Load a bobbin and your choice of thread in your embroidery machine, following the manufacturer's directions.

4. Follow the manufacturer's directions to load the backing material and the fabric onto which you are going to embroider into the hoop.

5. Load the hoop into the embroidery machine. Use the embroidery machine's menu to confirm that the needle is able to go to all points of the hoop without any obstructions, following the manufacturer's directions.

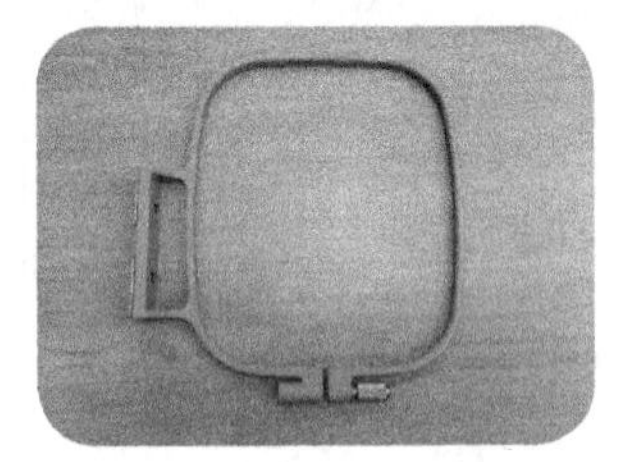

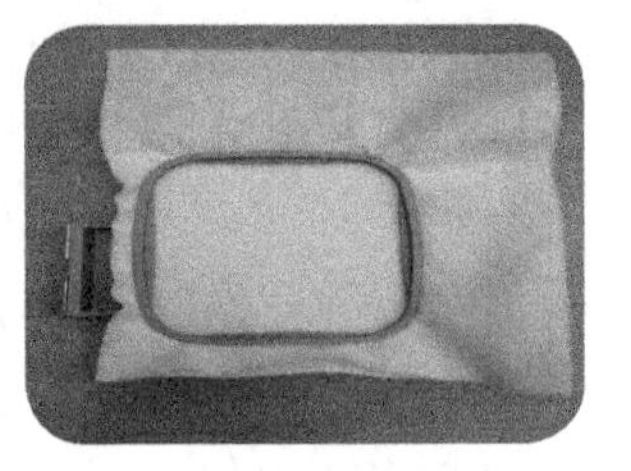

6. Once you have confirmed the machine is ready, embroider the design. Keep an eye on the machine while it works.

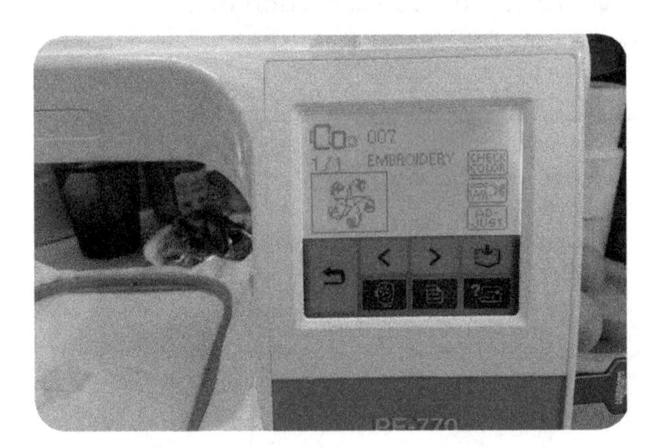

7. When the design is finished embroidering and before you remove the hoop, you may wish to embroider a border around the design if your machine has built-in capabilities. The Brother PE-770, for example, has a nice circular border that fit around the design. Consult your embroidery machine's manual for details.

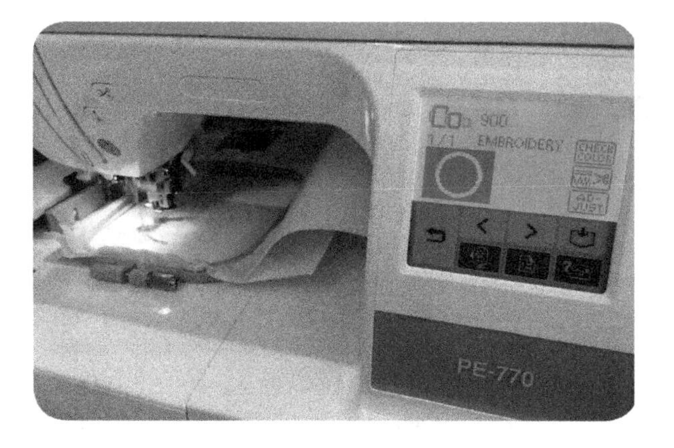

Taking It Further

If you are careful, you can even load backing and a piece of clothing into a hoop for embroidery directly onto the clothes. I embroidered a jeans jacket this way.

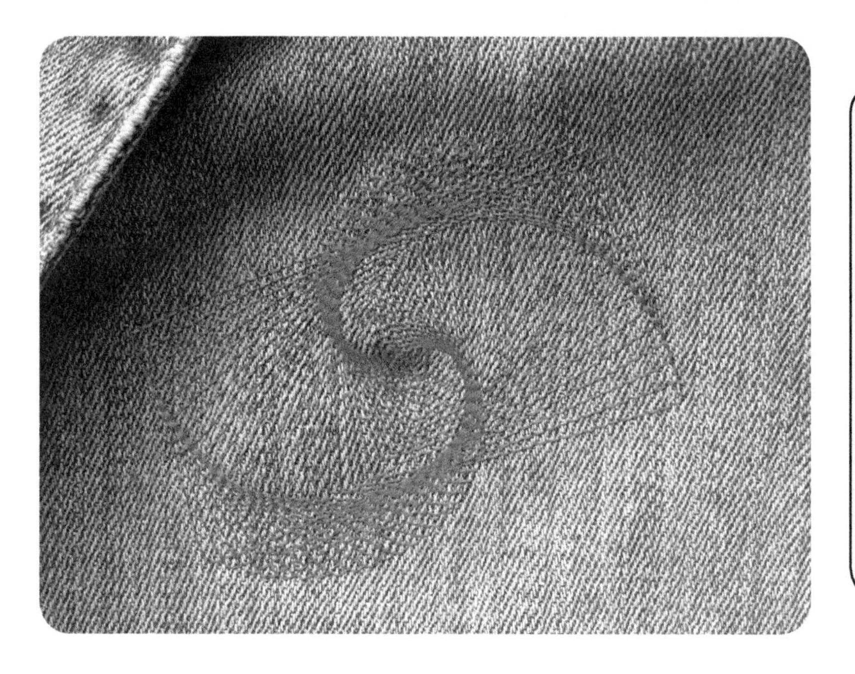

```
when clicked
reset
clear
set size to 0
point in direction 90
pen down
repeat 91
    move size steps by 10 steps
    turn ↻ 91 degrees
    move size / 2 steps by 10 steps
    turn ↻ 91 degrees
    change size by 5
pen up
```

Dylan Ryder showed me a trick of loading backing and a baseball hat into a hoop. The hat was held in place on the frame with binder clips. It's tricky—you can see that the design was embroidered off-center on the front of the hat.

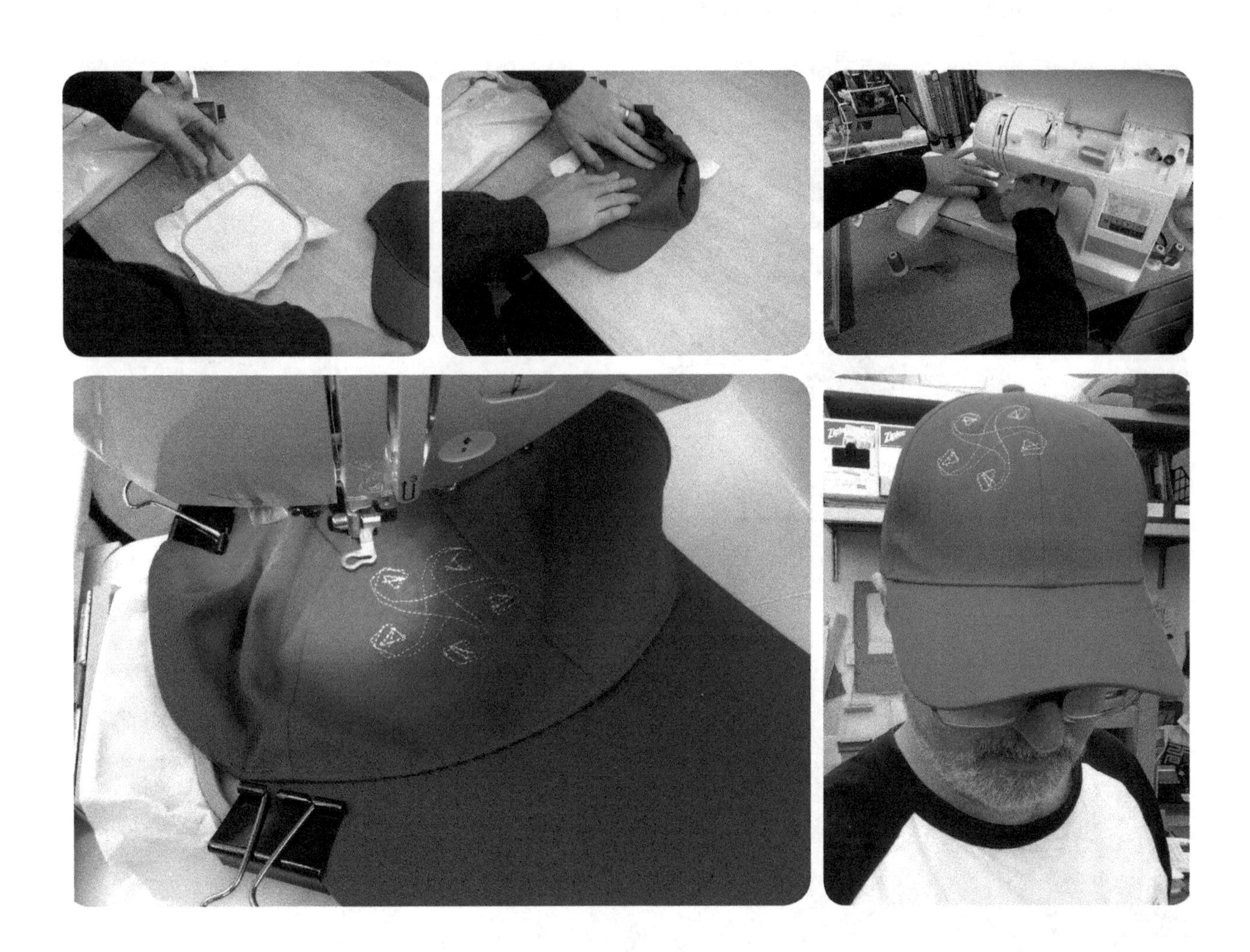

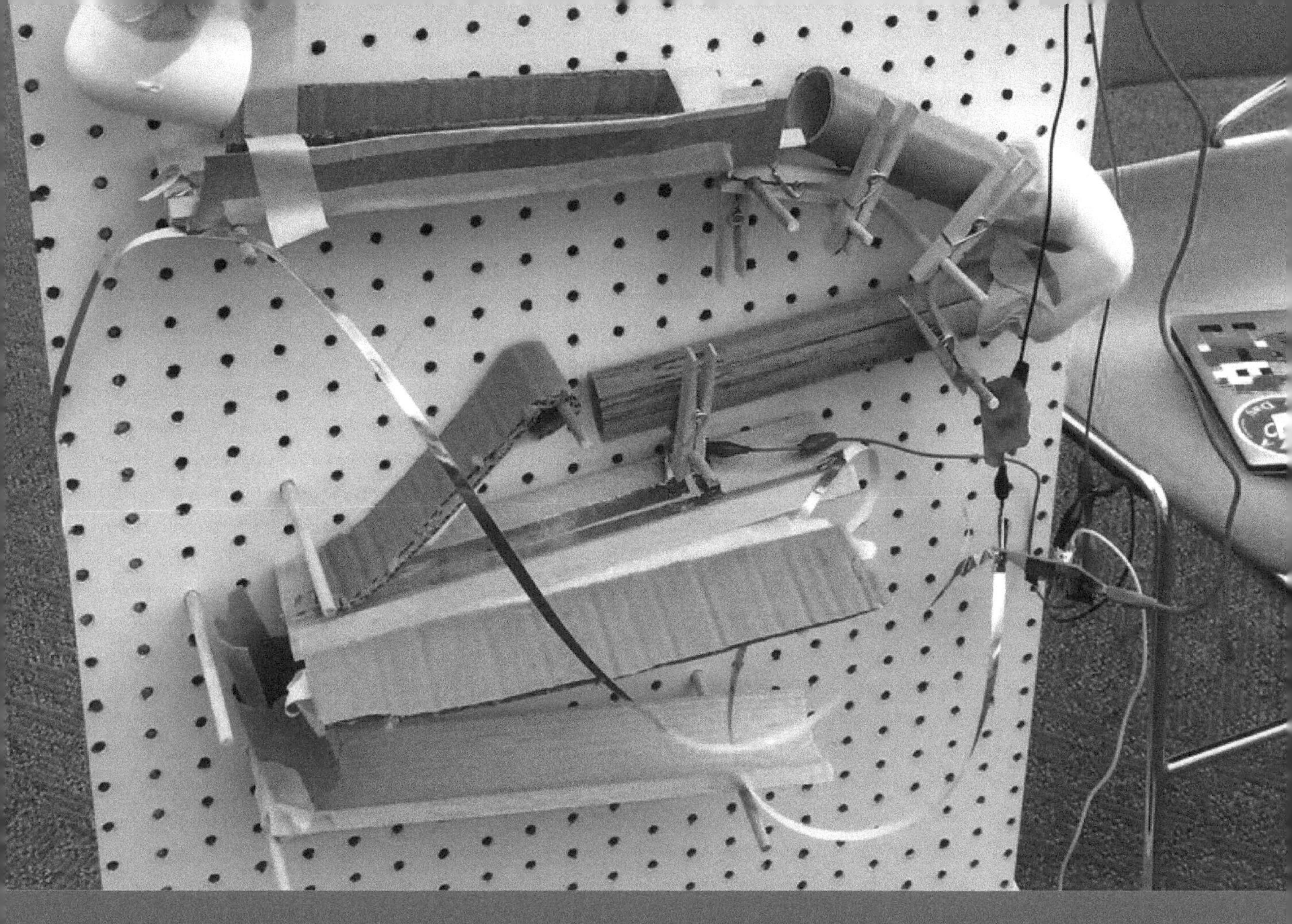

Multimedia Marble Machines

A Marble Machine is a wonderful tinkering project that encourages people to move a marble from the top of the peg board to the bottom using a series of ramps, tubes, funnels, and other obstacles and enhancements. People can create challenges such as trying to move the marble from the top to the bottom as slowly as possible. Often, after troubleshooting, adjusting, iterating, and resetting, the Marble Machine resembles a vertical Rube Goldberg machine!

One way to "upgrade" your Marble Machine to a multimedia tinkering toy is to add a little programming and circuitry to your Marble Machine. The Makey Makey is one of the simplest ways to do this. A Makey Makey detects completed circuits and triggers the computer to do some pre-programmed action, such as play a sound or animation. The key is adding sections to your Marble Machine that act as triggers, and using a metal ball. As the conductive metal ball rolls onto your triggers the circuit is completed and the Makey Makey responds.

This chapter does not lead you through constructing one Makey Marble Machine but rather guides you through some construction techniques and examples that you can remix and repurpose into your own Makey Marble Machine. Of course you can incorporate other technologies into your Marble Machine. This chapter has some ideas about using LEGO WeDo sensors and motors to make even more whimsical Marble Machines. For advanced Marble Machine enthusiasts, try out the suggestions for incorporating an Arduino, circuitry, and ScratchX or Snap4Arduino.

Materials

- Marble Machine pegboard: the design from The Exploratorium is easy to build and stands up to heavy use. Materials list and build instructions (exploratorium.edu/pie/downloads/Marble_Machines.pdf).
- An assortment of wooden ramps cut from molding into 12 and 4 inch lengths, ¼" wooden dowels cut to about 5 inch lengths, masking tape, marbles, and a small basket to catch and store your marbles. All of these materials are listed in The Exploratorium's build instructions.

- A selection of cardboard or bamboo tubes, wooden blocks with holes drilled through them to hang the blocks from dowels to act as stoppers, clothes pins, and PVC pipe angles are also very useful in Marble Machine construction.
- Conductive copper tape with conductive adhesive (a.co/7qMyYju) Note: conductive copper tape comes either with plain non-conductive adhesive or conductive adhesive. The conductive adhesive makes the project a bit easier, but is more expensive.
- Conductive marble: wrap a glass marble in a single layer of aluminum foil. Smooth out any wrinkles, test the marble on a ramp, and smooth as necessary. Alternately, a small metal ball bearing will work, too.
- Makey Makey
- Alligator Clips
- Cardboard
- Box cutter
- Cutting mat
- Pen or pencil
- LEGO WeDo 1 or 2 set
- Laptop to program WeDo 1, laptop, Chromebook, or tablet for WeDo 2
- Scratch or LEGO WeDo software installed on laptop, Chromebook, or tablet

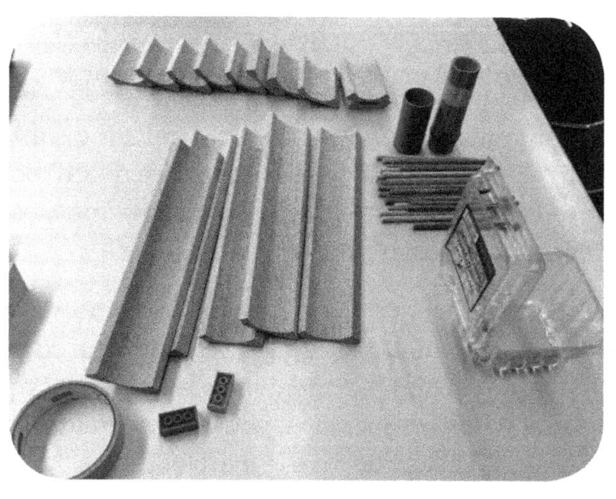

Strategies for Construction

Building a successful Marble Machine requires you to set up, test, and adjust the ramps, tubes, funnels, and whatever other paths and obstacles you devise for your marble. There are a few techniques to keep in mind if you plan to wire the Marble Machine ramps to use with a Makey Makey.

Copper Tape Triggers

One way to integrate circuitry into the Marble Machine is by using conductive copper tape with conductive adhesive to build triggers on your ramps. You can stick the conductive copper tape right to the wood or cardboard ramps you are using. Add two parallel strips of copper tape that do not touch, with tabs at one end made by folding the copper tape over on itself. Connect these tabs to the Makey Makey with alligator clips, one to ground and the other to one of the MaKey Makey inputs. When a conductive metal ball rolls over the two copper strips, it will close the circuit and trigger the MaKey MaKey.

Keep It Slow And Steady

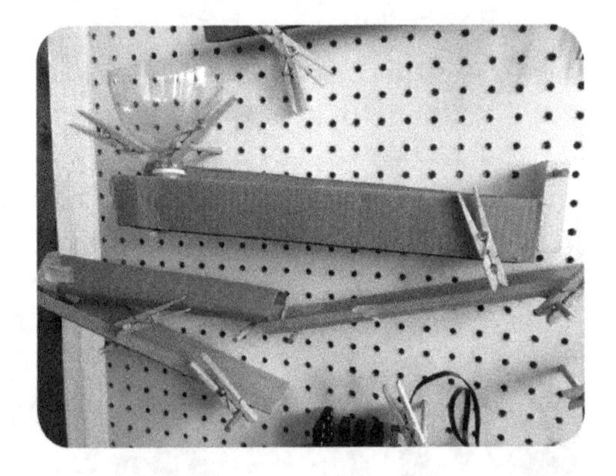

The Makey Makey needs a little time to detect a closed circuit and send a signal to the computer. If the marble rolls too quickly down the ramp you have wired for use with a conductive marble the Makey Makey will not have time to react. One technique to try to keep the marble rolling slowly down the ramps is to make some of the ramps level or even slightly titled upwards.

Remember, gravity is at work here! Ramps that are too tilted will cause the marble to move much too fast for the Makey Makey to sense a circuit being closed by a conductive marble.

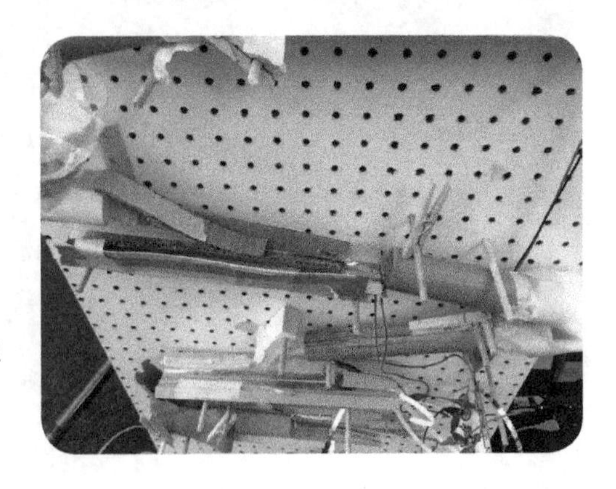

If you find that the marble is moving too fast across the ramps for the Makey Makey to detect the conductive marble closing the circuit between the two strips of conductive copper tape, you can try building a cardboard "finger" to slow down the marble. Tape a small piece of cardboard to a dowel and bend it so the finger lightly contacts the marble as it rolls by. Adjust your finger to slow down the marble to a speed detectable by the Makey Makey. Alternately, make the underside of the finger one half of your switch by adhering conductive copper tape to it as well as to the ramp below; when the marble makes contact with the finger it will close the circuit.

Another issue that can occur when using the conductive copper tape on the ramps is if the marble is moving too fast it might wobble and not make contact with both strips of conductive copper tape and close the circuit for the Makey Makey. Besides adjusting the pitch of the ramps to make the marble roll slower, you could also use cardboard to build "guardrails" that help keep the marble centered over the two strips of conductive copper tape.

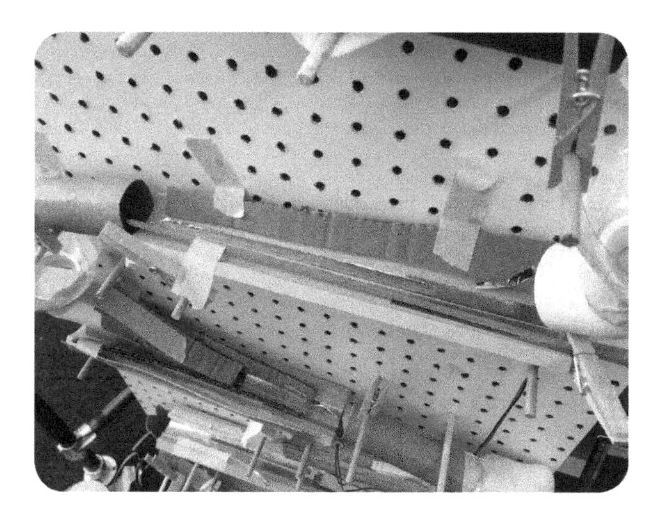

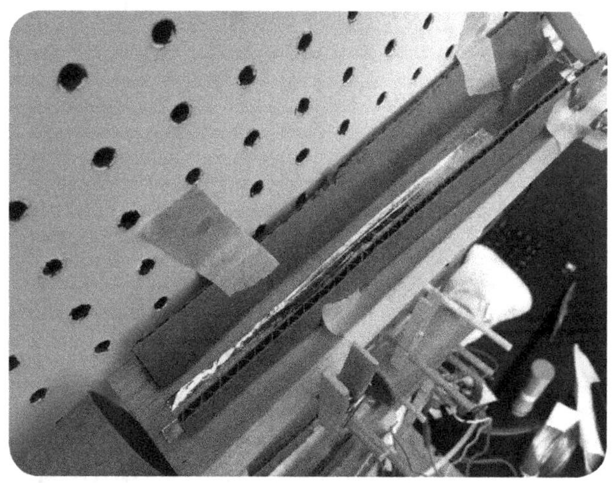

Adding Multimedia

You can connect one or more of these triggers to the Makey Makey and program Scratch to react to the key presses that the Makey Makey sends to the computer. Scratch can be programmed to make sounds, play music, or trigger animation on a connected computer.

A video of the first musical Marble Machine that I helped to build is here: vimeo.com/136980931.

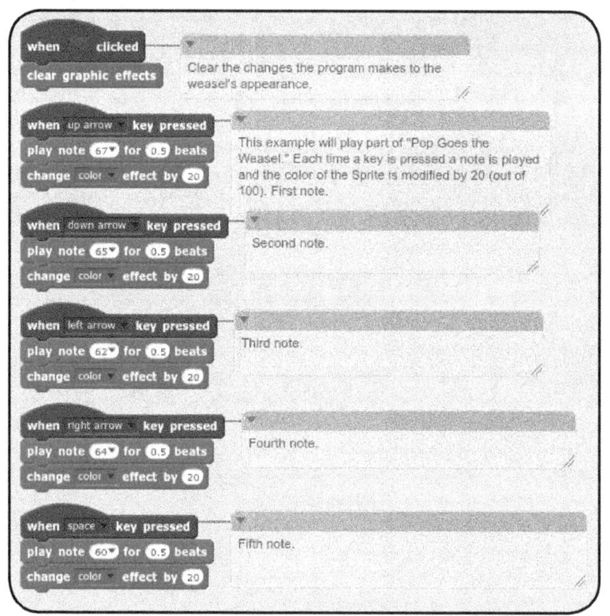

Other Triggers

An off the shelf switch can also be connected to the Makey Makey and integrated into your Marble Machine. A rocker switch, like the one pictured here, would work well at the end of a ramp where the marble would roll into the extended "finger" on the switch. You could make the target larger by hot gluing a small piece of cardboard onto the switch. Experiment with which pin on the switch you connect to the ground and one of the keys on the Makey Makey; you should need to connect only two of the pins to the Makey Makey.

Likewise, an oversized button placed at the end of a steep ramp could be hit and depressed by the marble. Experimentation is your best strategy!

LEGO WeDo

The LEGO WeDo set, both WeDo 1 and 2, include sensors and motors that can be integrated into your Marble Machine. These sensors can be used to trigger sounds or music to be played, just like the Makey Makey does with the "wired" ramps you created.

A distance sensor taped to a clothespin, clipped to a dowel, and aimed at the path of the marble on the ramp can accurately detect a rolling marble. In this example, programmed in the WeDo 2 software on a tablet, the motion sensor waits until it detects an object then plays a sound.

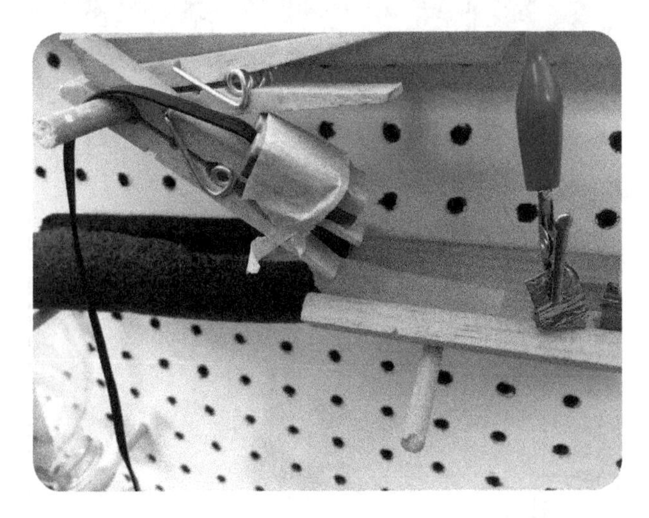

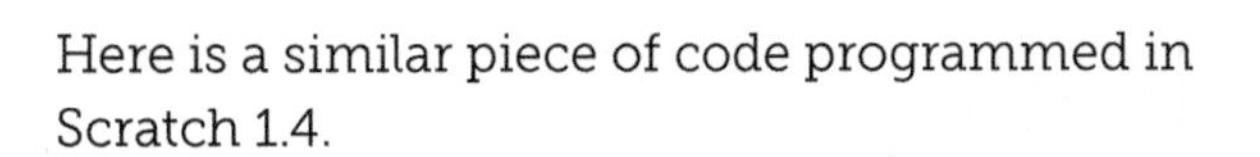

Here is a similar piece of code programmed in Scratch 1.4.

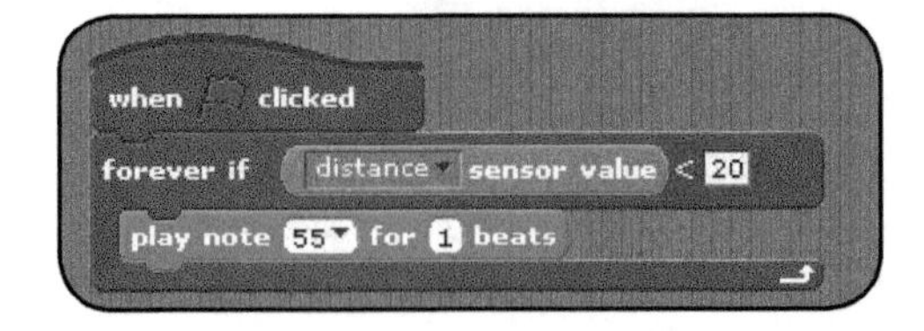

This Marble Machine featured a green and white XO (One Laptop Per Child) laptop running Scratch hanging on the pegboard. The WeDo 1 tilt sensor was placed at the end of a domino run that crashed into the sensor and tipped it over. The sensor rested on top of a stack of LEGO bricks that mimicked a domino in shape and size but had a pivot in its base. The marble would hit the domino and come to a stop while knocking over the row of dominoes and the tilt sensor.

When the Scratch program detected the tilt sensor changing orientation it started the LEGO motor. A "leg" attached to the motor kicked another marble down the next ramp to complete the Marble Machine run. A video is here: youtube.com/watch?v=eL9sHoVYFig.

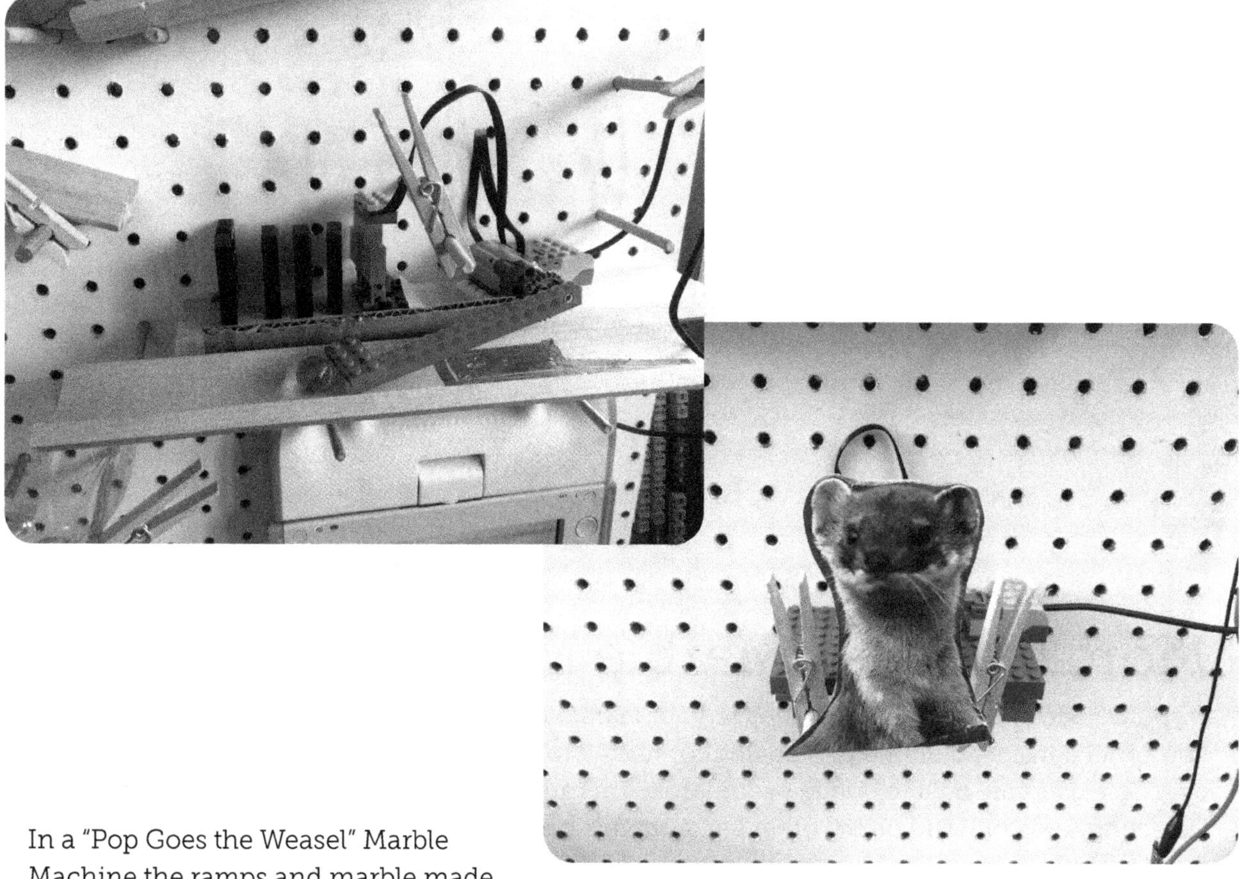

In a "Pop Goes the Weasel" Marble Machine the ramps and marble made the computer play the refrain of the song. At the end of the run, the marble triggered a motor with a weasel photo attached to turn 90 degrees counter clockwise for the "pop." You can see a video here: youtube.com/watch?v=gE9HIdItkzk.

More Complex Marble Machines

If you wish to add even more features to your Marble Machine and use more advanced tools, you could consider building photogates connected to an Arduino to "see" the marble passing by. These gates are built to bridge over marble runs and contain an LED on one leg and a Light Detecting Resistor (LDR) on the other leg. A marble passing through the gate causes the light to dim for a split second, but long enough for the LDR to trigger the Arduino.

For one project which I helped build with a sixth grade science teacher, we used photogates with ScratchX for Arduino (khanning.github.io/scratch-arduino-extension/) to make Scratch play notes and keep time as the marble rolled through a series of these gates. This section covers how to use these sensors with the experimental ScratchX extension system and how I built my photogates so you can remix and improve on the design.

Additional Materials

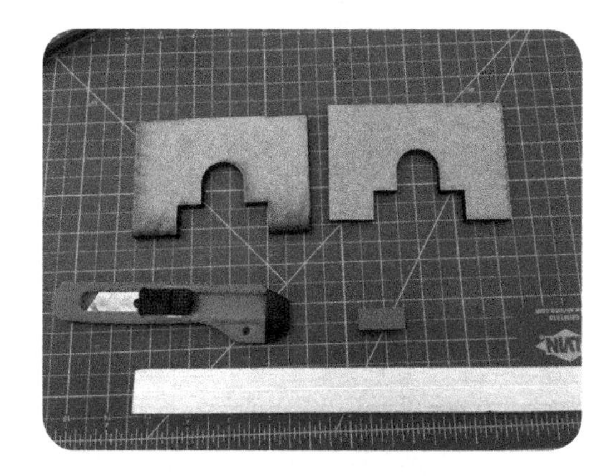

- The gate is laser cut from 1/4 inch MDF at the size you need. Additionally, a small strip of balsa wood is used as a spacer and a 3D printed bracket used for a set screw holder (thingiverse.com/thing:2442546).
- Arduino
- Breadboard, alligator clips, and connection wires
- Photocell (CdS photo resistor - as many as you have gates) adafruit.com/product/161
- LEDs (as many as you have gates) (adafruit.com/product/299)

Building the Photogates

1. Use the balsa wood as spacers between the two pieces of the gate.
2. Solder hookup wires to an LED and insulate the two "legs" from one another using electrician's tape. Do the same to a photocell.

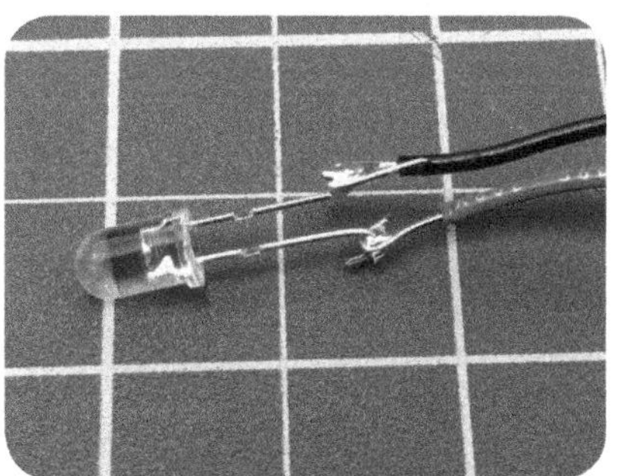

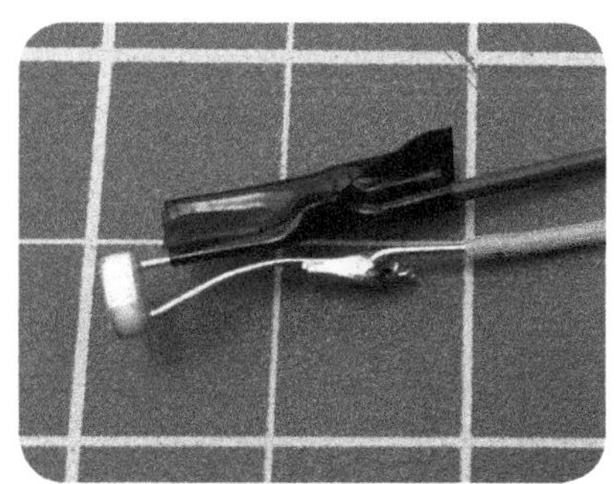

3. Use a short piece of drinking straw to position the LED and photocell so they point at one another across the gate. Hot glue the wires in place.
4. Add the other side of the gate by gluing it to the balsa.

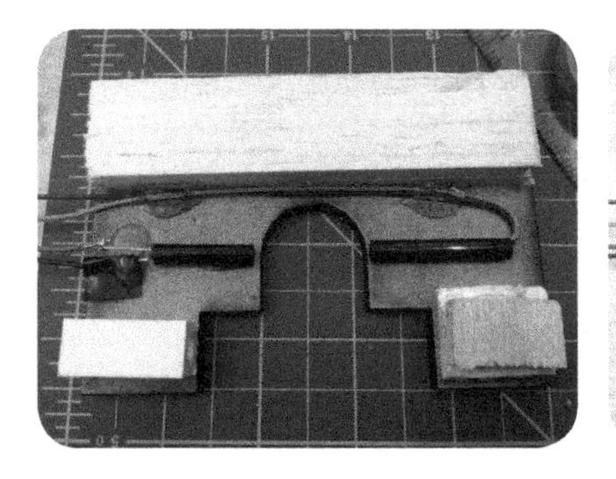

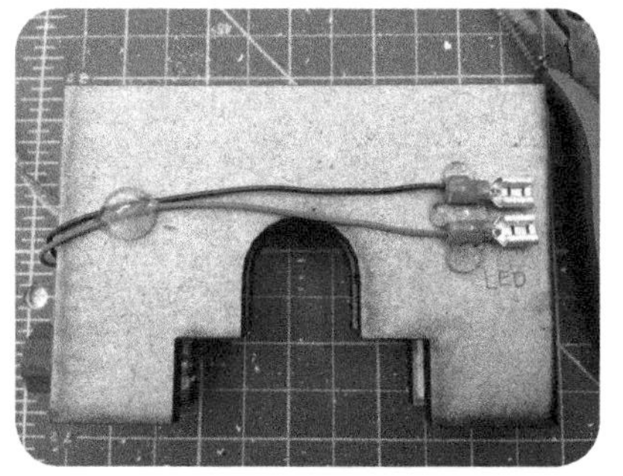

5. Route the wires to the top of the gate and label them. I added connectors to make it easy to connect the photogates to a long cat-5 cable I hacked to allow me to connect two gates per cable.
6. I also tapped the set screw holder and added a 3D printed thumb screw (thingiverse.com/make:336646) to make it easier to turn the set screw.

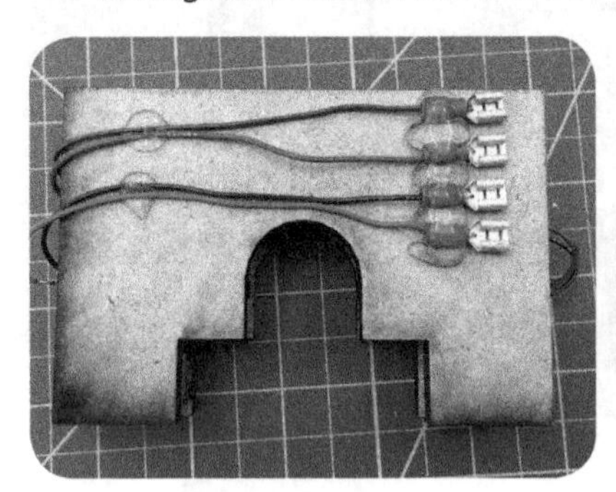
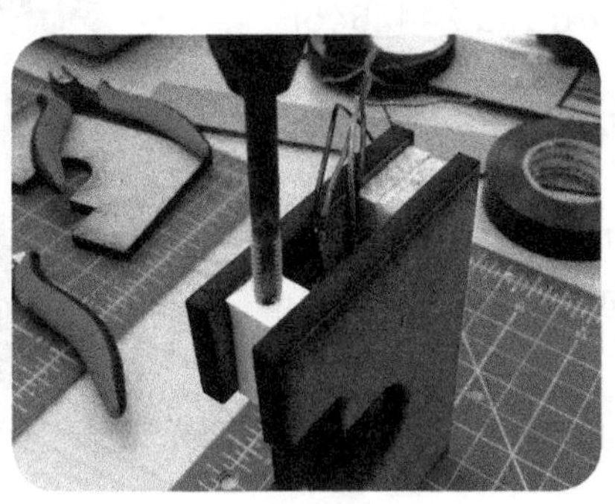

Once connected to the Arduino and ScratchX I was able to program the LEDs to turn on and read the photocell to monitor how bright the light was shining on the sensor.

The gates were arranged along a ramp and triggered as the marble passed through each gate and blocked the light. ScratchX recorded the times and played a note as each gate was triggered. As we experimented, we found that a glass marble just diffused the light and did not provide good sensor readings. We changed to a metal marble and found that it worked much better. This kind of tinkering experimentation is how learning happens!

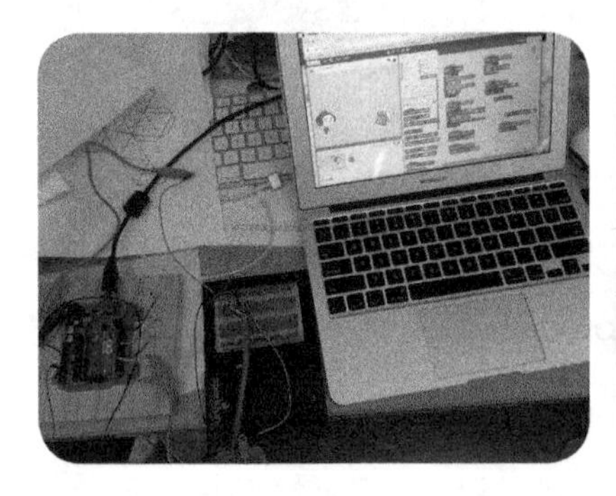
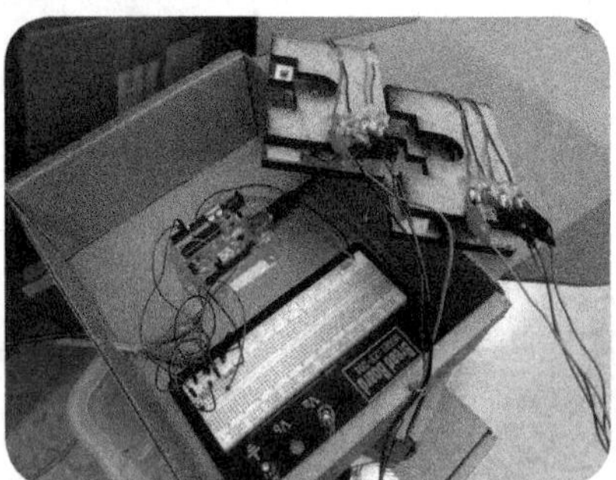

The Marble Machine starts as a blank canvas on which you can explore physics, electronics, music, art, and more. Completing a working Marble Machine takes patience, iteration, troubleshooting, time outs, and perseverance. By adding electronics, circuitry, and programming to your Marble Machine you can keep the floor relatively low but increase the ceiling to accommodate your wildest imagination!

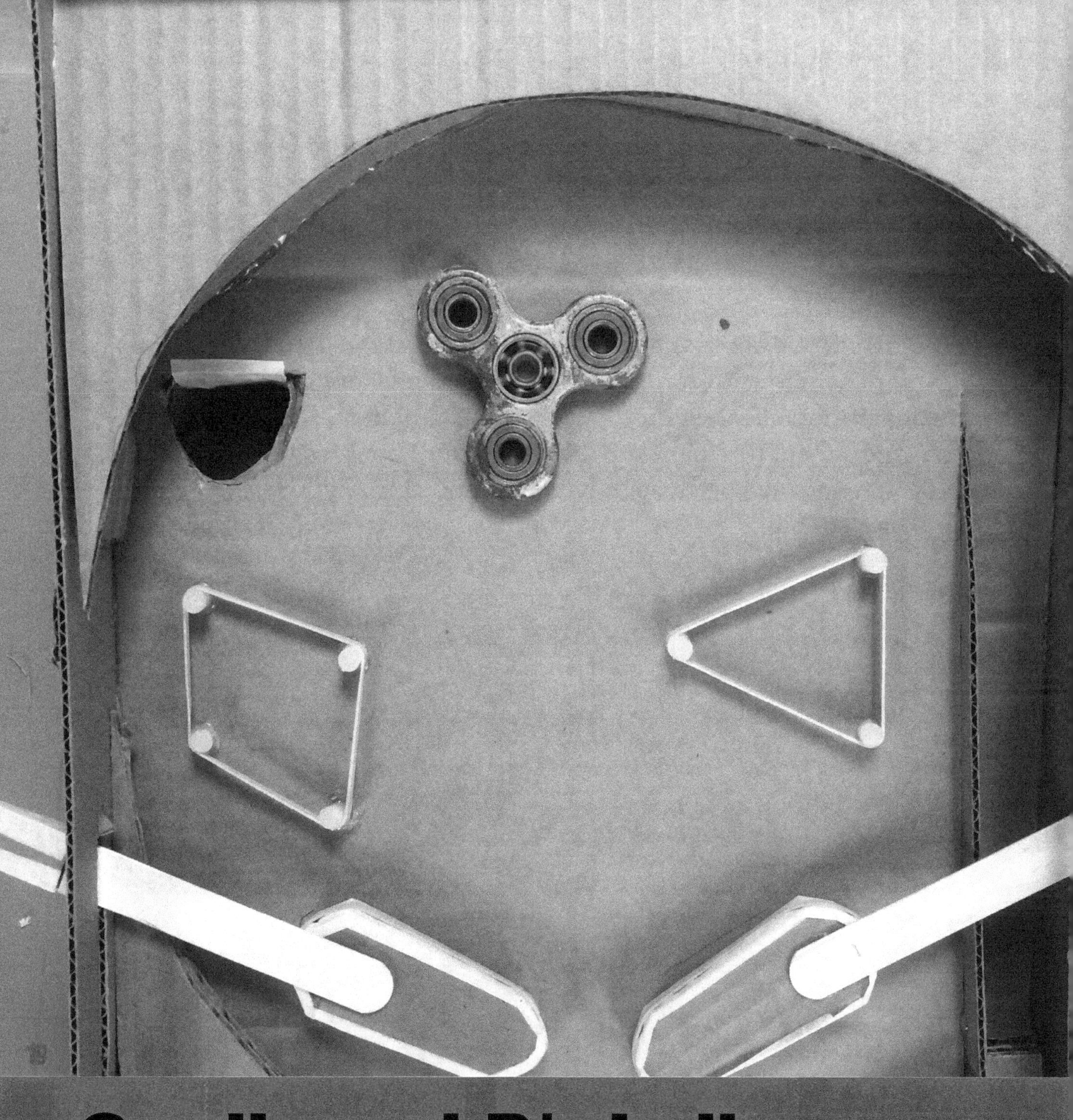

Cardboard Pinball Machines

At Constructing Modern Knowledge 2016 I saw a number of different pinball machines constructed. Some were simple mechanical constructions, while others used physical computing, integrating sensors, buttons, circuits, and Scratch. All of them were built in large part from cardboard.

What struck me about the pinball machines is how they used many different solutions to common engineering challenges in building the machines, like the flippers or the ball launcher. All of them were challenging but fun: hard fun, like Seymour Papert called projects that keep you engaged because they are tough, require you to think, plan, build, and debug.

Later on I decided to build my own cardboard pinball machine to see how I might engineer my way through some of the challenges. I applied some of the ideas I learned at Constructing Modern Knowledge and tried some of my own approaches, too.

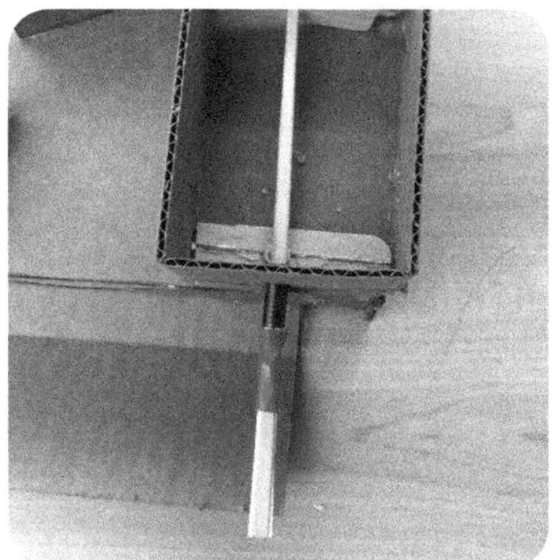

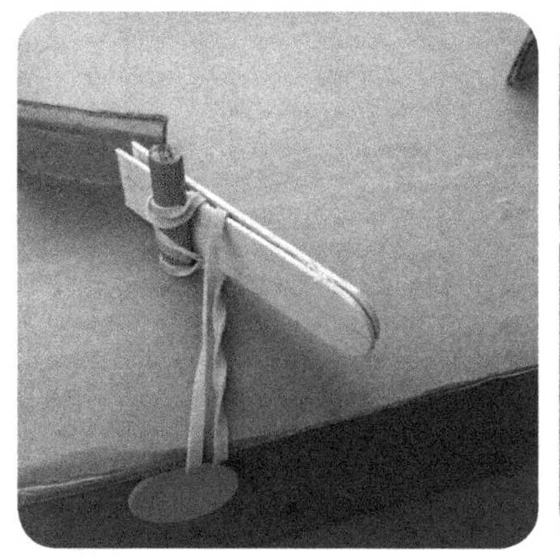

The purpose of this chapter is to acquaint you with some techniques and ideas of how to build a cardboard pinball machine. Iteration is fun and important in cardboard pinball machine construction. Like the Multimedia Marble Machine chapter, the tips and tricks here are meant to provide ideas, rather than a perfect recipe. The beauty and versatility of cardboard as a building material is on full display as well!

Materials

These are handy materials to have on hand in pinball machine construction.

- Cardboard
- Cutting mat
- Box cutter
- Pencil
- Ruler and measuring tapes
- Masking tape, painters tape
- Hot glue gun and glue sticks
- Mod Podge and brush for application
- Cardboard tube from a roll of paper towels
- Wide rubber bands
- 6 inch x ½″ tongue depressor or craft stick (wider and longer than a popsicle stick)
- Screws, nuts, bolts
- Screwdriver, awl, needlenose pliers, small hand tools
- Dremel Tool with 1/8″ flat milling bit (included in the standard Dremel kit as 194 High Speed Cutting Bit)
- Safety glasses
- Scrap piece of wood you can put underneath cardboard pieces you are milling
- Springs
- Chopsticks
- Marble

Constructing the Pinball Machine Body

The body of the pinball machine can be very simple or complicated, with hidden ball returns and other features. One consideration may be the experience of your students or workshop participants and how much time you have together. For example, in this photograph, I built a two level pinball machine with a ball return level under the main play surface.

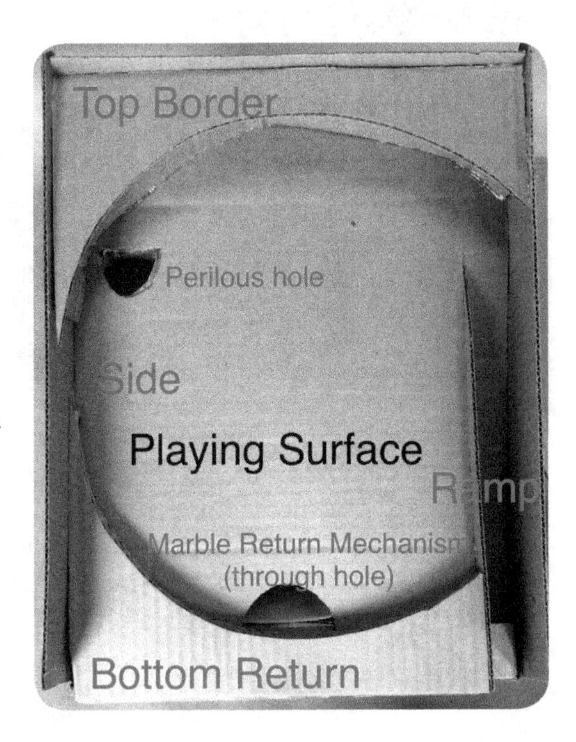

Playing Surface

The playing surface is the piece of cardboard the ball will roll down, bounce around, and encounter obstacles.

- Construct the playing surface out of an existing box. The flat surface will be where the ball rolls and the walls will keep the ball in play, and also provide a place to attach flippers or a ball launcher.
- Use a box with a large flat surface so the ball doesn't have to roll over flaps or seams.
- Use sturdy cardboard so you can attach obstacles and flippers or cut holes in it without losing stability.
- The pitch of the playing surface affects the game's difficulty because the pitch establishes the marble's speed down the playing surface. A steep pitch will make the marble roll quickly down the playing surface, while a more shallow pitch slows down the marble's speed. You can adjust the pitch later to change game play.

Flippers

This is a simple flipper design that is solid and holds up to use but is open for improvement. You can experiment with the rubber bands used to pull the flippers back into position: if the flippers are too hard for you to move, try a thinner (or longer) rubber band.

1. Place a small piece of scrap cardboard near the bottom of the play surface. Use your pencil to sketch and refine a flipper shape that is appropriately large for your playing surface.

2. Cut out the flipper using your knife: you will use this piece as a template.

3. Use the template and a pencil to draw eight copies of the flipper on a piece of cardboard. Use the template to help you cut out eight copies of the flipper shape.

4. Combine the cut flippers into two stacks of four pieces. Hot glue the flipper shapes together into stacks of four.

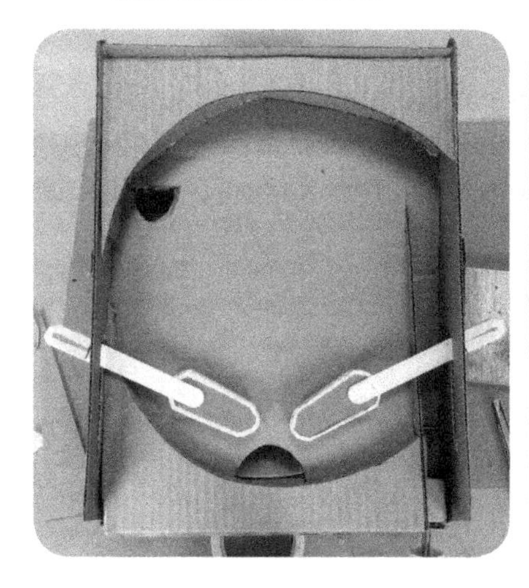

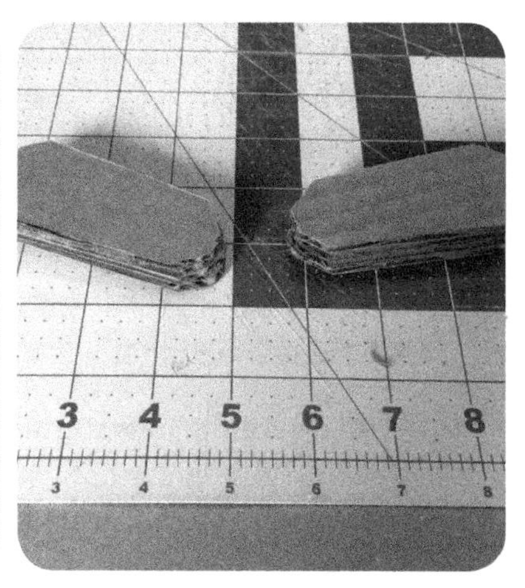

5. Wrap the edges of the flippers with masking tape.

6. Use your scissors or knife to cut slits in the tape around the curves. This makes it easier to cleanly fold down the tape.

7. Place a rubber band around the flipper to provide a little bounce when the marble makes contact.

8. Apply an even coat of Mod Podge to the cardboard top and bottom of the flippers and the folded tape. Do not apply to the rubber band areas: you will use them to balance the flippers while they dry.

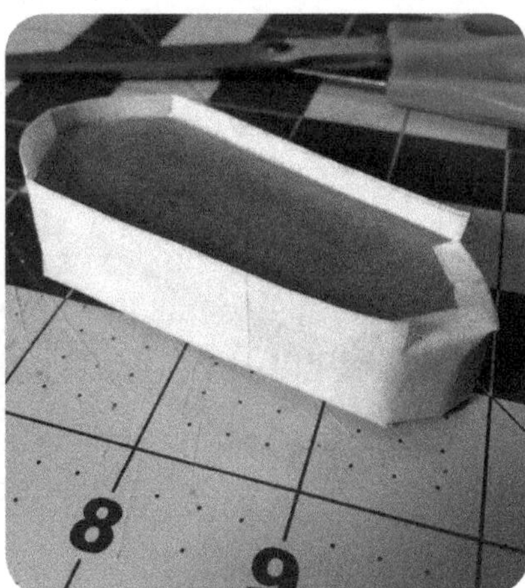

Flipper Arms

Once the flippers are dry position them on the playing surface. Trace the flippers with a pencil onto the playing surface so you can remember their location. You are going to build the arms that allow you to move the flippers.

1. Hold a craft stick so one end is touching the flipper and the other end touches one side of the box, as shown. Mark the side of the box slightly above where the stick meets the side of the box. Additionally, make a mark perpendicular to the stick to show the maximum "up" travel range the stick and flipper will have.

2. Adjust the flipper's position and repeat the process to mark the maximum "down" position the stick and flipper will have.

3. Turn the pinball machine on its side. Use your ruler and knife to cut out a small rectangle through which the craft stick will pass.

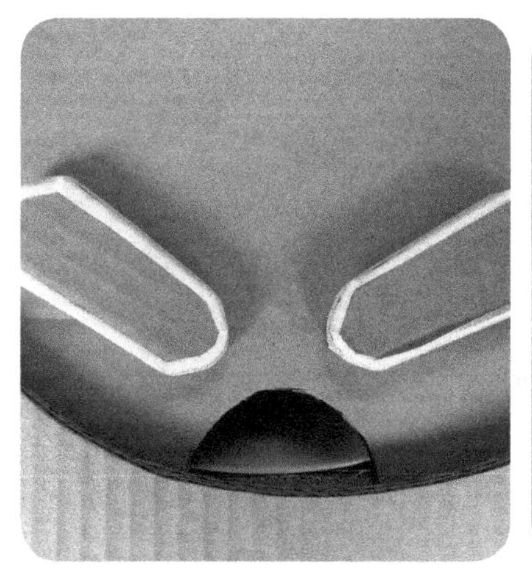

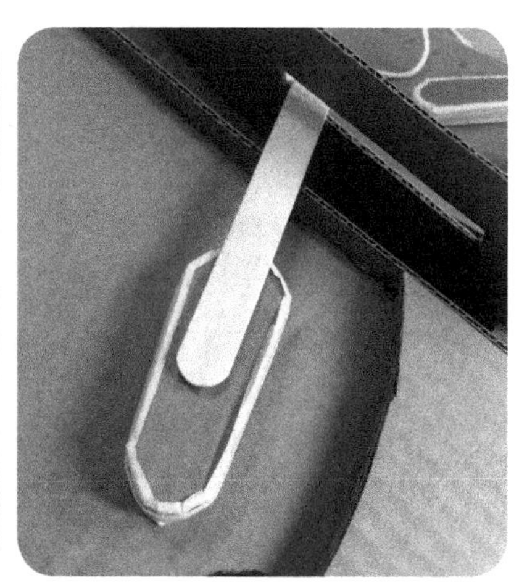

4. Repeat steps one and two for the other flipper and side.
5. Test the clearance of the craft stick and flipper through the side holes.

Attaching Flippers to Playing Surface

We will use crown bolts with plastic anchors to securely attach the flippers to the playing surface. This connection should hold up to repeated use but is simple to construct.

1. Mark one of the flippers with the location for the screw, towards one end of the flipper.
2. Put on your safety glasses and following the manufacturer's directions load the 194 high speed cutting bit into the Dremel tool. Place a piece of scrap wood underneath the flipper you are going to mill. Hold on to the flipper while you mill straight through all four pieces of cardboard. Run the Dremel at the high speed setting.
3. Place the flipper back on the playing surface in the spot you outlined. Use the flipper as a guide to mill a hole through the playing surface in the correct position for the flipper. Hold on to the flipper while you use the Dremel tool.

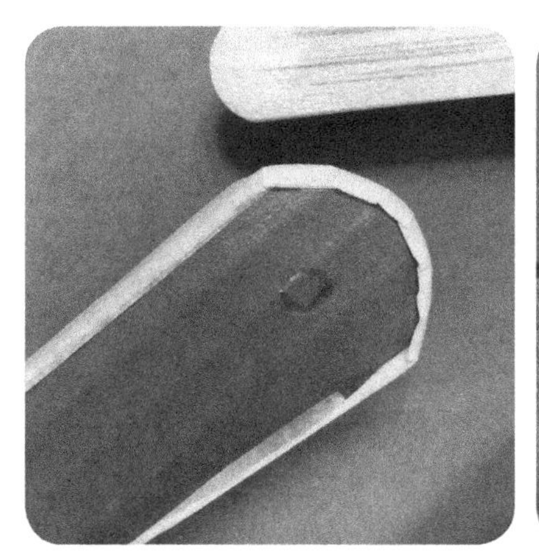

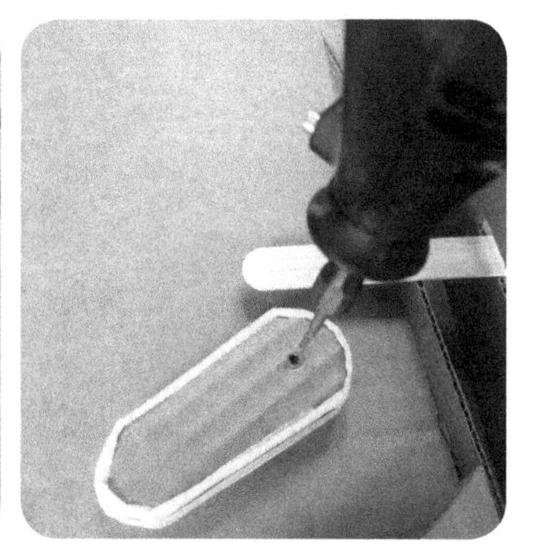

4. Stack the flipper with the hole on top of the other flipper, oriented in the same direction, on top of the scrap wood. Hold onto both flippers. Use the top flipper as a guide to help you mill a hole through the other flipper. Remove the top flipper once you go through the first layer of cardboard on the bottom flipper, then finish milling a hole through the bottom flipper. You can unplug the Dremel and take off your safety glasses now.
5. Use the second flipper and the directions in step 3 to drill a second flipper hole in the playing surface. Insert a plastic anchor into one of the playing surface flipper holes. Gently push it until its last ring sits slightly above the cardboard.
6. Remove the plastic anchor, apply hot glue towards the top of the anchor, and reinsert the anchor into the hole.

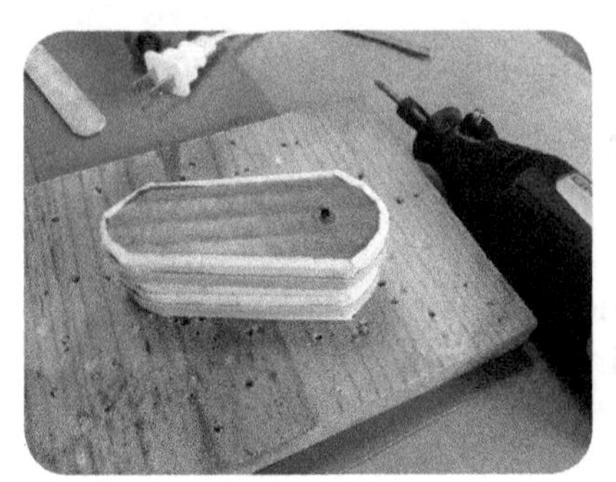
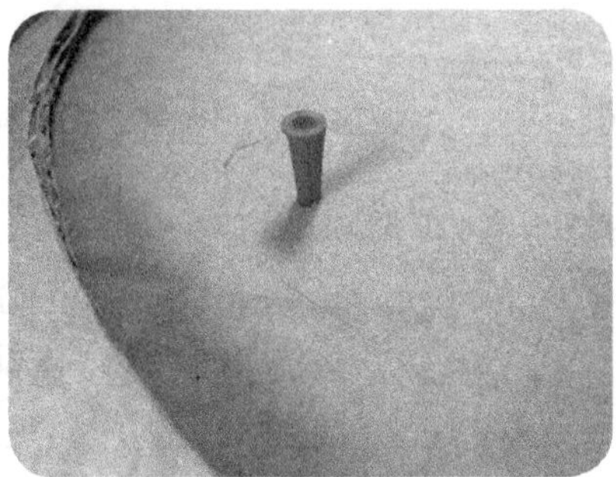
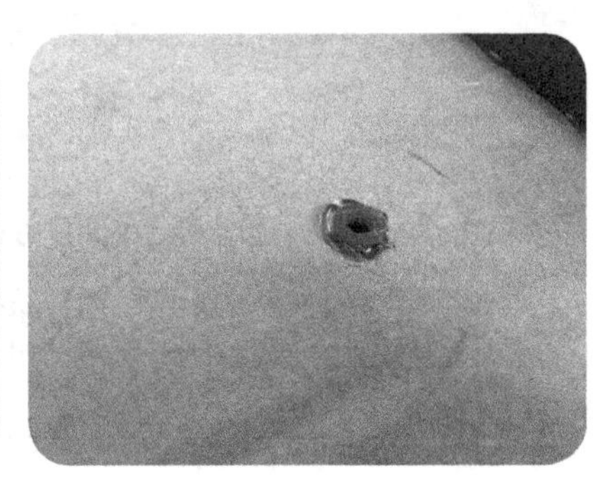

7. Repeat steps 6 and 7 for the second plastic anchor.
8. Use the screwdriver to turn the screw through all four layers of cardboard on a flipper. Try to depress the first layer of cardboard so the crown head is recessed.
9. Use the screwdriver to drive the screws into the plastic anchors. Be careful not to overtighten, as you want the screws to remain in place but the flippers to rotate around the screw without gripping (you are essentially "stripping" the cardboard so the screw acts as a pin).
10. Test and make sure the flippers move both up and down with minimal effort.

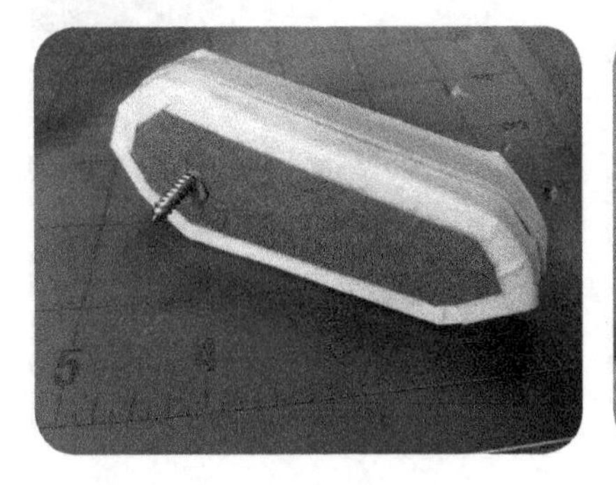
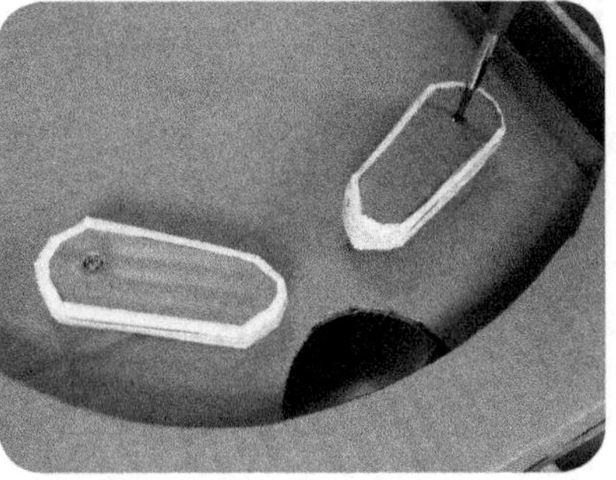
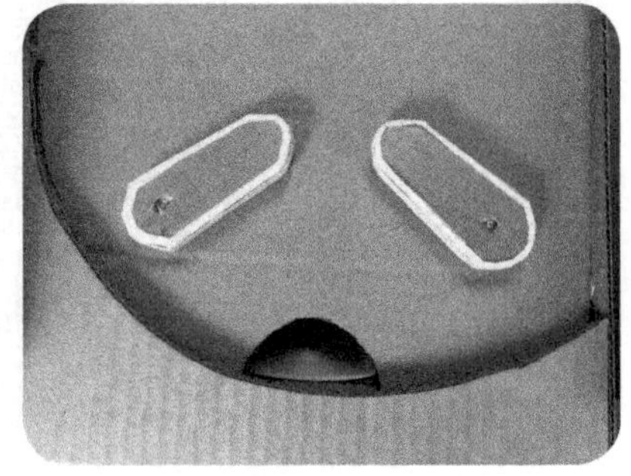

Attaching the Flipper Handles

1. Insert the flipper handle through the slot cut in the side of the pinball machine side. You want for there to be a small section of craft stick visible as you will connect a rubber band to it. Mark the position of the flipper on the craft stick.
2. Apply hot glue to the underside of the flipper handle using the mark as a guide. Press and hold it to the flipper until the glue hardens.
3. Repeat steps 1 and 2 for the other flipper handle.
4. Cut a strip of cardboard about 2 centimeters wide and 10 centimeters long. Drape a rubber band around one of the flipper arms. Working from the side of the pinball machine, slide the cardboard rectangle through the rubber band and pull it back until the rubber band is taut but not stretched. Mark where the rubber band sits on the cardboard: this is the area that you will glue, leaving an area without glue for the rubber band.
5. Apply hot glue to the cardboard strip and affix the strip to the side of the pinball machine
6. Repeat steps 4 and 5 for the other flipper handle.

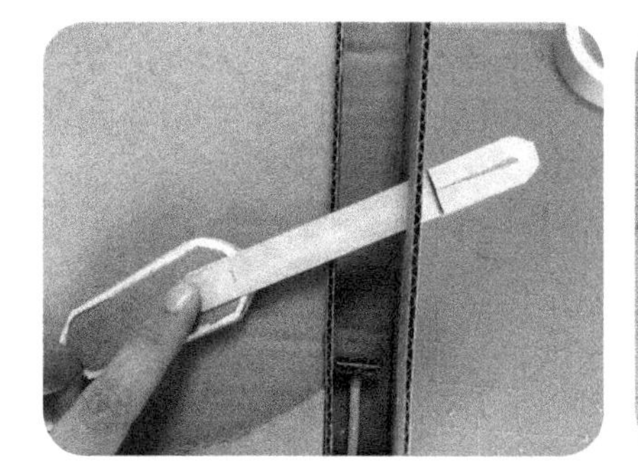
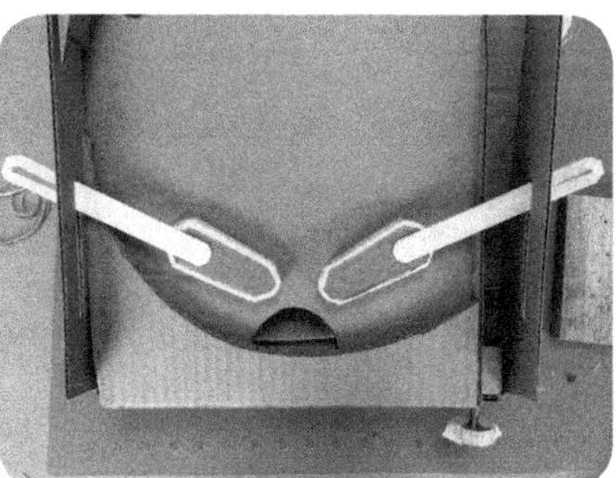
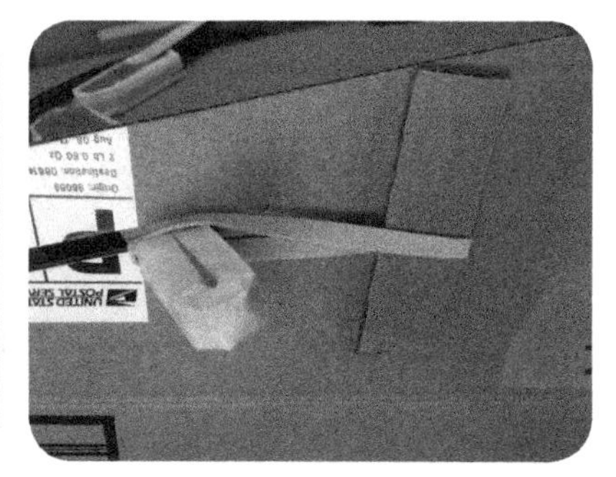

Congratulations! The flippers are in place and ready to use. Use your fingers to give them a quick flick, pulling them toward you to raise them and letting them go to snap them back into position. If you find they are too stiff try using a narrower or longer rubber band. You will have to cut the hot glue on the top of the cardboard rectangle but that is easy to patch after you swap in a different rubber band.

Constructing the Marble Launcher

Your cardboard pinball machine can have a pull back / release marble launcher. How fast the marble launches up the ramp depends on how far back the launcher is pulled. This example uses a spring on a chopstick for an easy build.

1. Scavenge through an assortment of springs until you locate the one that is compressed until

you pull it apart and that fits around your chopstick. We are going to call the chopstick the launcher rod.

2. Apply a spot of hot glue to the end of the launcher rod and the spring to hold it in place. You may need to gently twirl the launcher rod and blow on the hot glue to keep it in place while the glue dries and hardens.

3. Use the launcher rod as a guide to help you position where you will mill a hole through back of the cardboard box. Put on your safety glasses. Use the Dremel tool to mill the hole through the cardboard in the appropriate position.

4. Insert the launcher rod through the hole and wiggle it around slightly to loosen the hole. You want the launcher rod to pass easily through the hole.

5. With the launcher rod inserted so the spring touches the cardboard box, use the needlenose pliers to bend the ring at the end of the spring so it sits flat against the box. Remove the launcher rod and spring from the cardboard box.

6. Cut a small square of cardboard. Use the knife and the tapered end of the chopstick to make a hole in the center through which the launcher rod fits. Cut a slit from one edge of the square to the hole.

7. Insert the launcher rod, spring side first, through the hole and slit in the cardboard. Leave the ring you bent in step 7 on the other side of the cardboard square than the rest of the spring.

8. Insert the launcher rod back through the hole you put in the cabinet. The spring ring will be between the cardboard box and the small square around the launcher rod.

9. Brace the cardboard square against the back of the pinball machine cabinet. Pull back on the launcher rod. Mark a spot on the launcher rod where the maximum pull distance is.

10. Place the chopstick on the scrap wood. Cut the launcher rod at the spot you marked using the knife and a rolling motion of the launcher rod.

11. Use the ruler to measure the width of the marble launcher ramp. This ramp is about 2 centimeters wide.

12. Cut a rectangle of cardboard slightly less wide than the ramp and twice as long as this measurement (2 cm x 4 cm were my measurements). Score the cardboard in the middle of the long side and fold the cardboard into a square.

13. Cut a V notch in one side of the rectangle, wide enough to fit the launcher rod.

14. Insert the launcher rod into the box. Apply a swirl of hot glue to the side of the cardboard square facing the box.

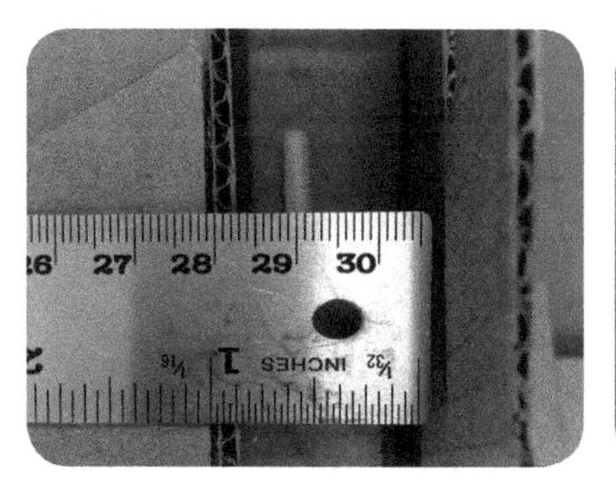

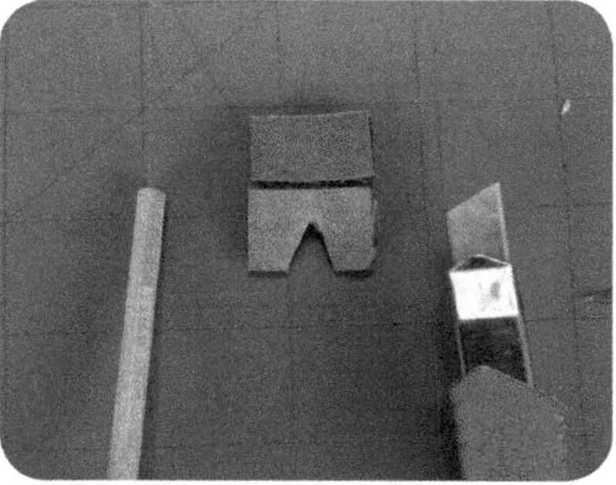

15. Affix the square to the rear of the box.

16. Apply a small amount of hot glue to the V notch in the folded cardboard square. Stick the square on the end of the launcher rod, small side of the V facing down, in the marble launch ramp. Hold the square in place until the glue cools and sets. Watch for any dripping hot glue that might threaten to secure the launcher rod to the marble launch ramp.

17. A bottle cap makes a good launcher rod handle. Attach with hot glue.

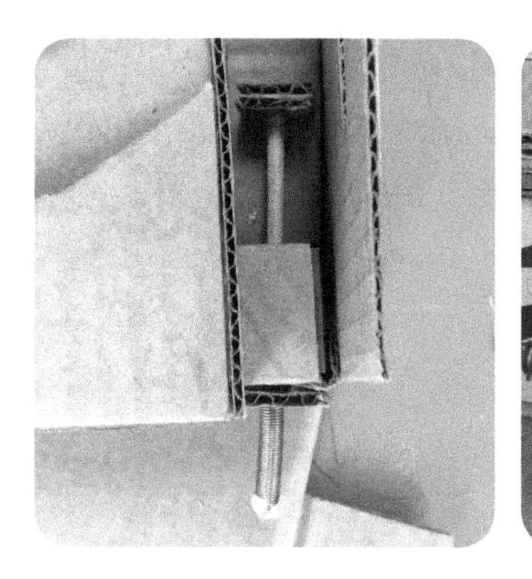

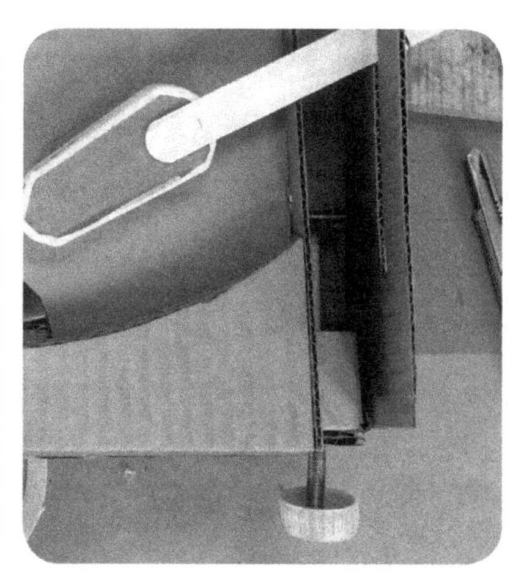

Obstacles

The cardboard pinball machine needs a couple of obstacles on the playing surface to make it more difficult and to keep the ball from being consistently launched back into the bottom return. Here are some ideas for two popular kinds of obstacles: bumpers and a spinner.

Bumpers

1. Cut pieces of hot glue sticks to half an inch long.
2. Use the tip of the hot glue gun to slightly melt one end of the short hot glue stick, then add a little hot glue from the hot glue gun before sticking them to the playing surface in a triangle or polygon pattern.
3. Stretch a rubber band around the glue sticks on the playing surface. Repeat steps 1 through 3 for as many bumpers as you want on the board.

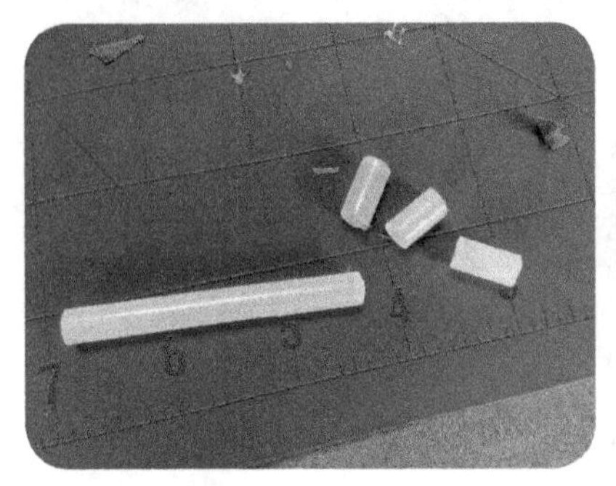

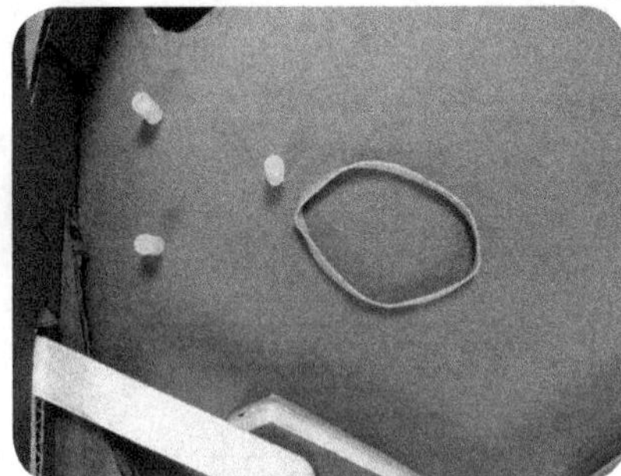

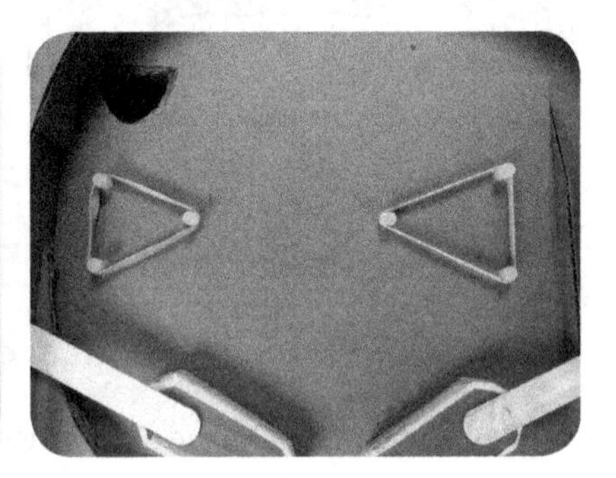

Spinner

1. Position a fidget spinner on the playing surface. I held mine down with a square of masking tape. When the marble hits the spinner the spinner starts moving and throws the marble off in an unpredictable direction. Taping the spinner down allows you to find the best position for it on the playing surface once you start playing. After fine tuning you can glue down the fidget spinner for more durability.

Troubleshooting

If the marble rolls down the playing surface too fast, a half inch roll of masking tape placed under the marble return slot tilts the machine slightly and changes the pitch of the playing surface.

You may want to reinforce the sides of the pinball machine where your hands are going to rest and pull against when they are using the flippers.

Have fun, play fairly, and share your improvements on Twitter and your blogs!

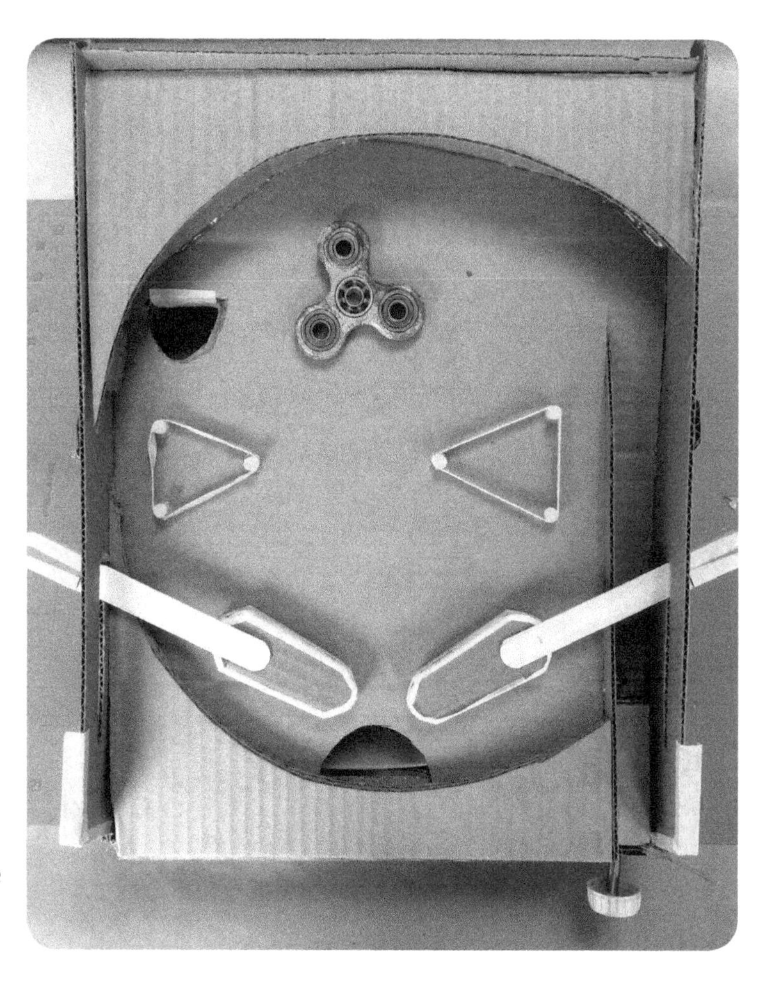

Play!

Your pinball machine is ready to play!

Make note of which parts work well and which parts could use improvement. Play with this cardboard pinball machine while you collect a list of improvements to make in your next iteration. I think I would improve my next iteration by making the playing surface twice as long!

Taking it Further

You can add multimedia elements to your pinball machine by using a metal ball and triggering animations or sounds using Scratch. See the Marble Machine chapter for ideas and techniques. You could also use WeDo sensors, or even sensors connected to an Arduino to control lights and sound effects.

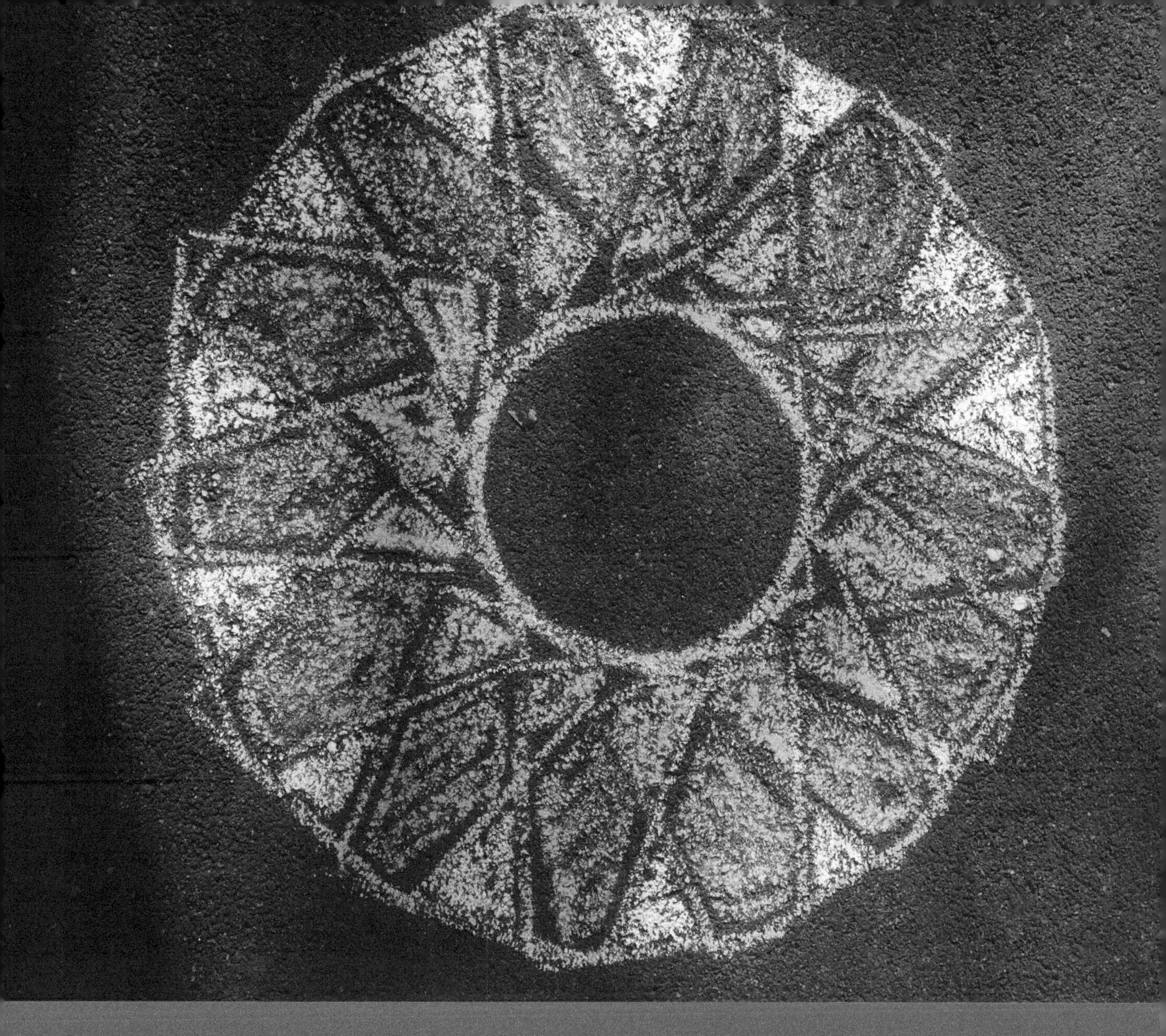

Sidewalk Chalk Geometric Art

You can explore the beauty and art of geometry away from the computer by creating cardboard templates to trace with sidewalk chalk. Whether you build one shape template or many different shapes, you will have fun tracing, overlapping, rotating, and creating using geometry.

Materials

- Cardboard that you save from the recycling bin (if you can find double-ply cardboard boxes, like those used to ship heavy materials, the template will hold up better to extended use)
- Box cutter
- T-square
- Ruler
- Pencil
- Hot glue gun and hot glue sticks
- Duct tape
- Cutting pad or other surface on which you can cut the cardboard
- Scissors
- Sidewalk chalk in a variety of colors

Optional

- 3D printed rivet x 2 (thingiverse.com/thing:551396)
- Awl, like on a Leatherman tool

Cardboard Fabrication Basics

First, let us discuss some cardboard fabrication basics. If you can find double-ply cardboard (I got mine from my neighbor who bought a new lawnmower), use it. This cardboard will hold up to being pushed into the asphalt or concrete. It is a bit tougher to cut but worth it for its durability.

Second, when you cut cardboard, remember the words of my friend Joseph Schott: you are not a gorilla. Do not try to cut through the cardboard in a single slice. Instead, make careful,

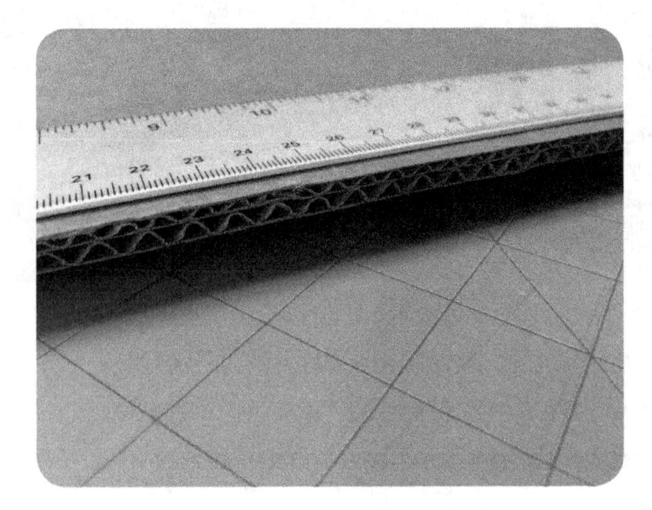

measured slices using a ruler or T-square as a guide. It might take four cuts to get through a standard piece of cardboard; it took me eight or so to cut through the double-ply cardboard.

Use a sharp blade to make cutting easier and safer. Always cut in a pulling motion, not a pushing motion, with all your fingers and body well clear of the blade. By making shallow, precise cuts with the blade you avoid the risk of accidentally losing control of the blade by pulling on it too hard and cutting yourself or marring your material.

Templates

As you look through these directions to build templates for the sidewalk chalk geometry, you may wonder if it's really worth it to spend so much time on them. However, sturdy templates with handles and smooth edges that hold up to repeated use will prove their worth.

Square Template

1. Use the flaps from a large box to build your square template. You can use two of the already cut edges to insure straight lines on two of the sides of your template.

2. Use a T-square on the long edge of the flap to measure the length of the box flap to the box body. In the photo below, the flap is about 32 centimeters long, or about 12 ½ inches.

3. Using that measurement as a guide, mark the box flap width at this length with a pencil then cut two squares from the cardboard flap using the T-square as a guide for your knife.

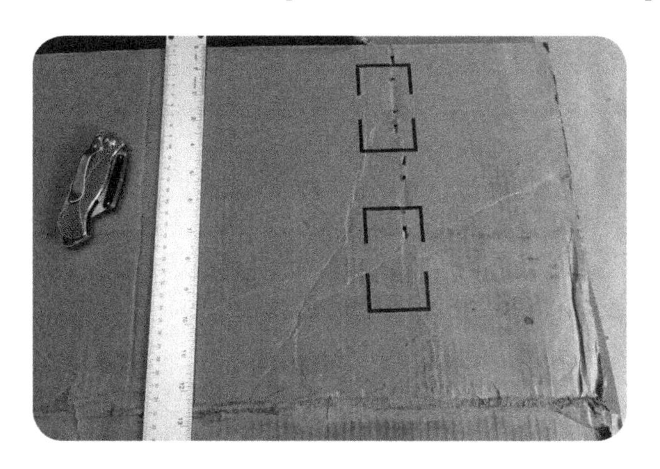

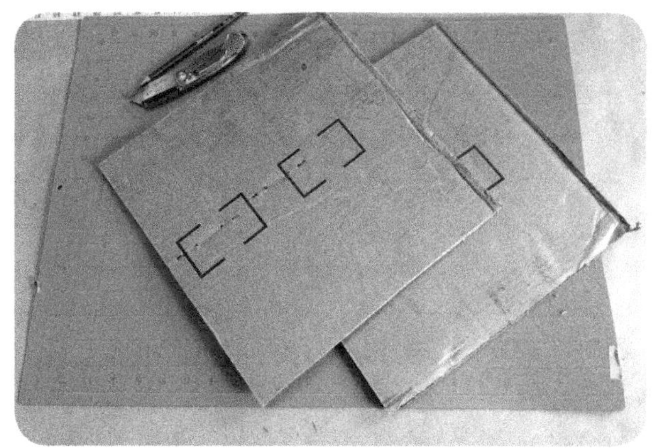

Circle Template

1. A large mixing bowl or bucket lid make excellent templates to create another template. Place the bowl upside down on the cardboard and use your pencil to trace the circumference. This way if the bowl moves while you are cutting the cardboard you can place the bowl back in the right position.
2. With the bowl still in place start gently slicing through the cardboard. Remember, you are not trying to cut through all the layers in a single pass.
3. Cut two circles from the cardboard.

Fabricating a Handle

Adding a handle to the back of the template allows you to easily lift up the template and reposition it without getting your fingers in the chalk and smearing the design.

1. Cut a strip of cardboard 10 centimeters wide and a little longer than your template is wide.
2. Use a ruler to find the center of the cardboard strip at 5 centimeters. Mark the top and bottom of the cardboard strip, lay the ruler along the center of the cardboard, and gently score the cardboard with the blade, making sure not to cut all the way through the cardboard.
3. Fold the cardboard with the scored edge on the outside of the fold.

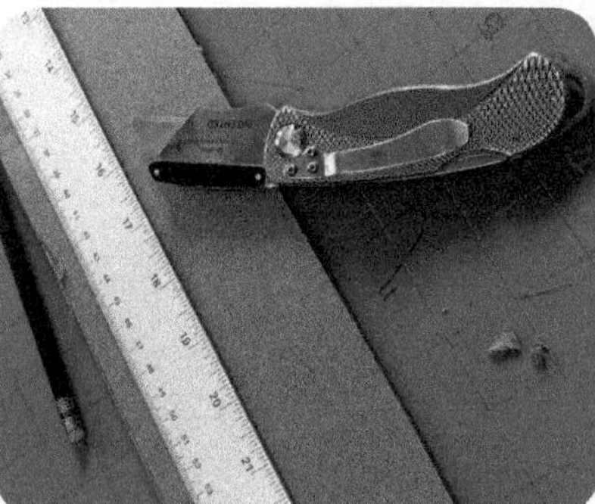

4. Cut a piece of duct tape as long as your cardboard strip. Affix the duct tape to one side of the cardboard, leaving enough tape to fold over and adhere to the other side of the strip.

5. Flatten the cardboard at the scored cut and affix the duct tape to the other side. Trim any excess duct tape. You now have a sturdy handle to connect to your template.

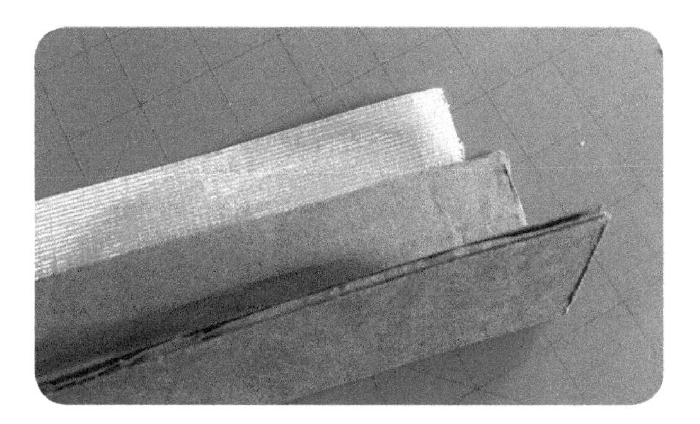

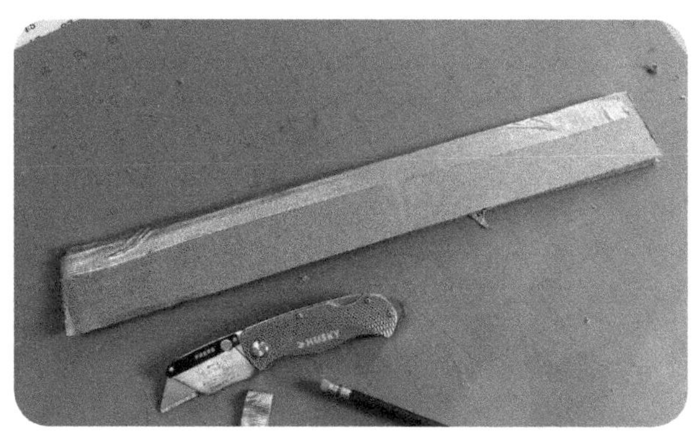

Connecting the Handle

If you do not want to use the 3D printed rivets, use these directions to connect your handle to your template. For this tutorial the square template is used but you can connect the handle using this technique to any shape template.

1. Using the handle you created as a guide, trace the edge of the handle on the template, about 5 centimeters from the edge. Repeat on the other side, parallel to the mark you made previously.

2. Using the marks as your guide, use the box cutter to cut out slots in the template for the handle.

3. Insert the ends of the handle through the slots. Use a ruler to help bend about 5 centimeters from each end of the handle so it begins to lay flat against the template.

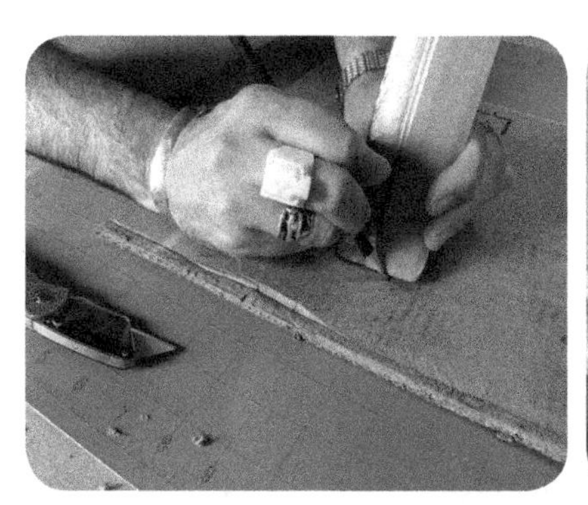

4. Cut two strips of cardboard that cover the ends of the handles, as show. Use hot glue to affix the strips to the template, holding the handle in place.

5. In order to even out the surface so you can affix the other side of the cardboard template, hot glue a couple of strips of cardboard cut to approximately the same size as the first two and glue them perpendicular to the first two strips, as shown.

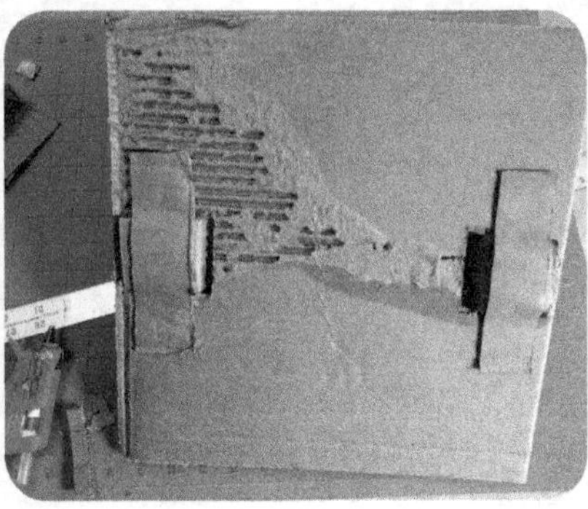
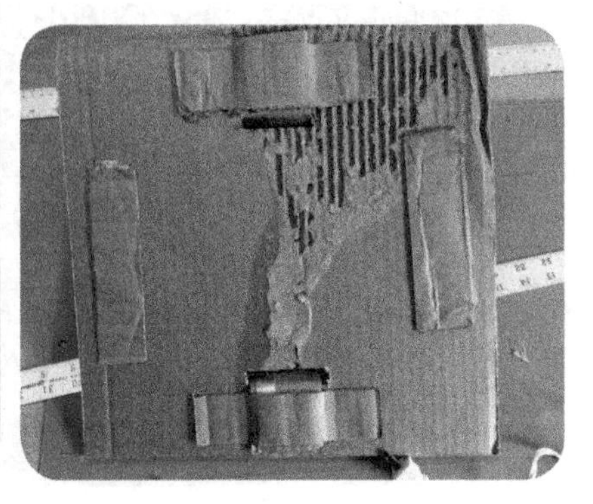

The handle is now installed. You can skip the next section if you are not attaching a handle using 3D printed rivets to other templates.

Connecting the Handle with 3D Printed Rivets

Using 3D printed rivets (thingiverse.com/thing:551396) is an easy and effective way of attaching the handle to the cardboard template. Typically I 3D print the rivet in ABS; the material's flexibility seems to make the rivet work better.

1. 3D print two copies of the rivet.

2. Position the handle where you want it on the cardboard template. It is helpful to bend the cardboard so it is more flexible: holding the handle at each end and running the handle over the edge of a table is a handy way to "break" the cardboard and make it more flexible. Use an awl to puncture a hole through the handle and one of the two cardboard shape templates you previously cut.

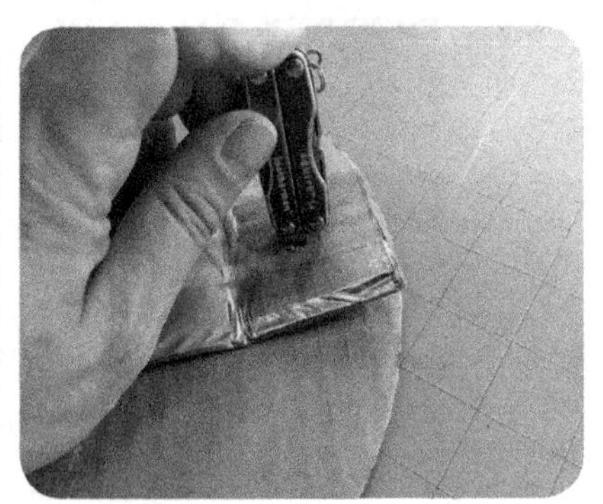

3. Use a pencil to widen the holes in the cardboard so the rivet will fit through the hole.
4. Push the rivet through the hole in the handle, making sure the head of the rivet rests against the cardboard; trim excess cardboard as necessary. Next, push the rivet through the template, connecting the handle to the shape. Flip over the cardboard template and handle and place the other half of the rivet on the rivet, pushing it down as far as it will go.
5. Cover the rivet and washer with a small piece of duct tape.
6. Repeat steps 2 through 5 for the other side of the handle and template.

You have now attached the handle to the template with the 3D printed rivets. Take care not to use the template as a shield; the rivets can pull out if put under too much stress. You can always pop them back into the holes in the cardboard to snap the handle back into place.

Attaching the Bottom of the Cardboard Template

To wrap up the building part of this project, you will attach the other side of the template to the first using duct tape.

1. If you are attaching two squares or polygons with straight edges together, cut a piece of duct tape as long as the side is wide. Using the same technique from the handle assembly, attach the duct tape to one side of the template and fold it over the edge with the second

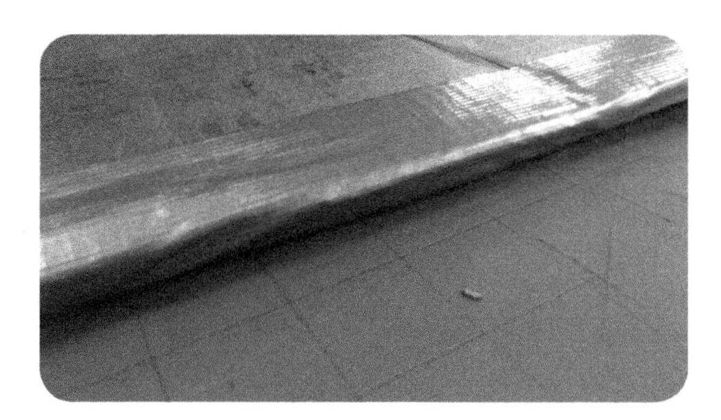

piece of cardboard stacked on top of the first. Affix the duct tape to the other side.

2. Repeat on the opposite side of the template, then finish with side three and four. Your template is ready to use!

If your template is circular, use these directions to attach the two halves.

1. Use two small strips of duct tape affixed opposite one another to attach the two halves of the circle template together.
2. Cut a strip of duct tape that is long enough to reach from one of the small pieces on the template to the other. Adhere the duct tape to the edge of the circle, making sure to press the two cardboard templates together as you stick the tape to the edge.
3. Use scissors to slice the duct tape from its edge to the cardboard every 2.5 centimeters or so along the strip you affixed to the template.

4. Working one flap at a time, pull and press down the duct tape on the face of the template. You should overlap each flap with the one next to it, as shown.
5. Repeat with the other side of the circle template. You have a complete cardboard template ready to use!

Creating Art with Your Templates

Now is the fun part: creating art using the templates. Get your sidewalk chalk and some friends and head outdoors to play with math!

Find a nice expanse of concrete or asphalt free of traffic. A driveway or footpath works great If you are making your art in a public place make sure there are no ordinances against using chalk there. My students and I have used the templates and chalk in Riverside Park in New York City with no complaint from park employees or police. Check the rules for your public spaces before using chalk there.

1. You can create complex patterns using a single cardboard template shape. Trace the perimeter of the shape with the sidewalk chalk. Again, you are not a gorilla: if you trace lightly the chalk will transfer to the pavement or asphalt and you will not wear down your chalk as fast.
2. Rotate (and move, if you want) your template a little, and trace it again.
3. Continuing to rotate and trace the shape quickly creates complex geometric designs.
4. Stand back and admire your math!

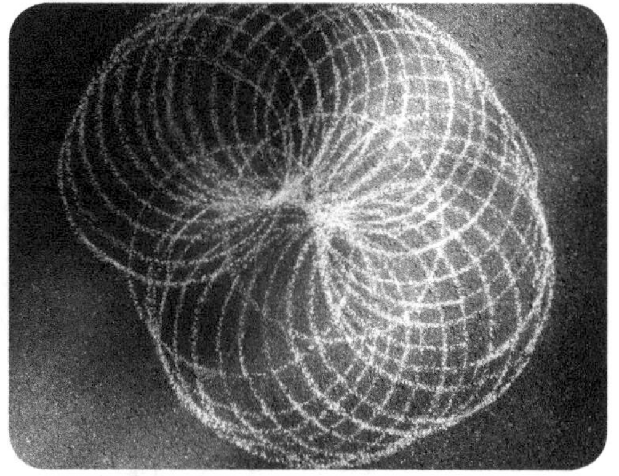

Taking It Further

Once you create geometric designs, start taking it further by shading some of the design with a different color chalk. You can fill in the secondary shapes that are created by overlapping shapes, for example. Try combining different cardboard template shapes in a single design. Fill in secondary shapes with different color chalks.

Make sure to photograph your work: your art will quickly disappear on the bottom of people's shoes and down the drain when it next rains.

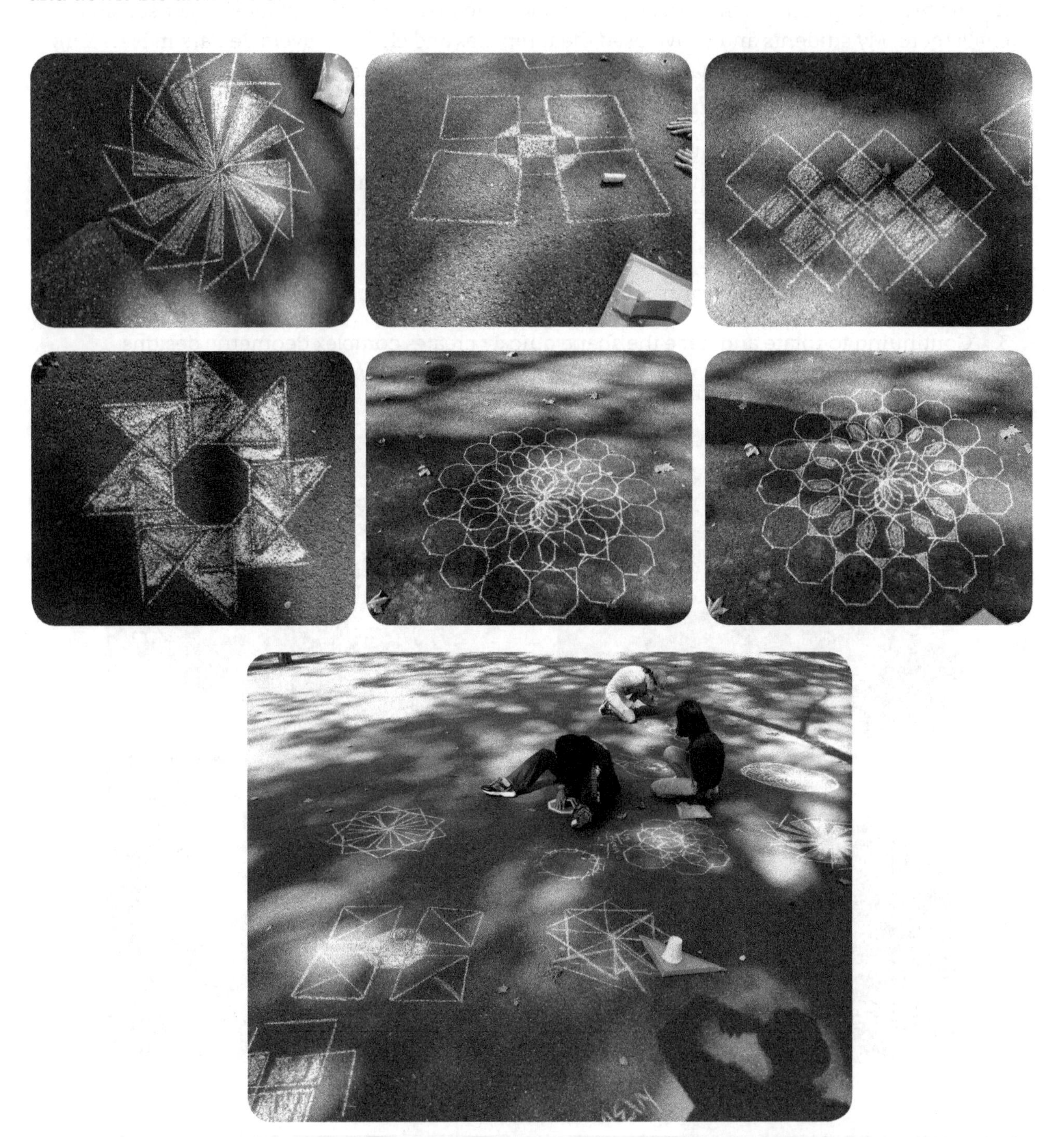

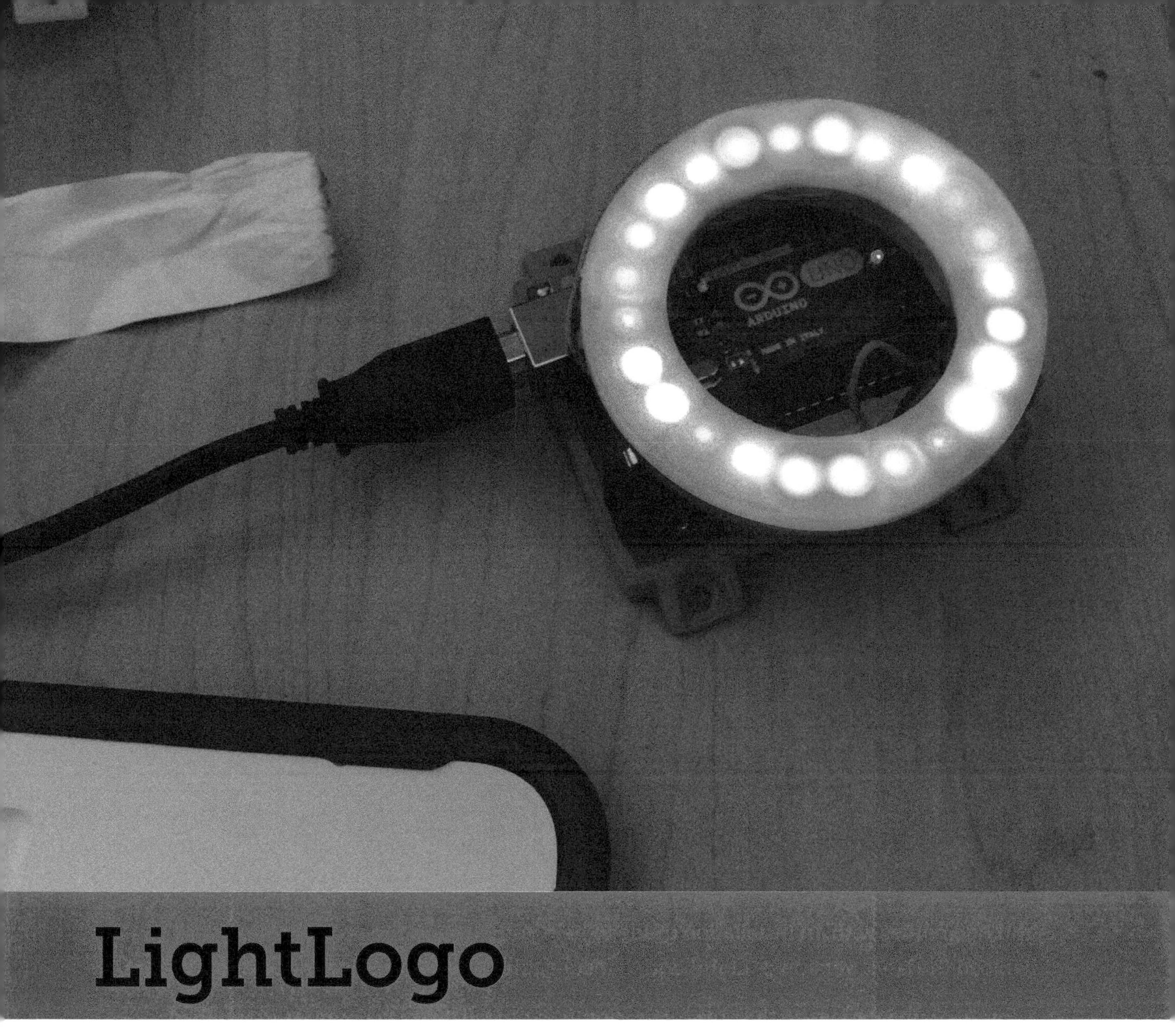

LightLogo

LightLogo is a Logo programming environment in which you program patterns of light using an Arduino and a 24 pixel NeoPixel ring. Brian Silverman, its creator, explains, "LightLogo can make interesting patterns emerge with ten lines of code." LightLogo brings Logo programming to a microcontroller, providing the opportunity to program light without the overhead of the Arduino IDE. This chapter leads you through setting up LightLogo, an introduction to programming and uploading your procedures, as well as some projects to extend LightLogo.

Materials

- Arduino Uno or Sparkfun Red Board (sparkfun.com/products/13975)
- Current FTDI driver for your system (ftdichip.com/Drivers/VCP.htm)
- LightLogo: current version (playfulinvention.com/lightlogo)
- 24 pixel NeoPixel ring (sparkfun.com/products/12665)
- 20 gauge solid core hookup wire
- 10K resistor
- Wire strippers/cutters
- Soldering iron and solder
- USB Cable to connect your microcontroller to your computer

Optional

- 3D printer and filament
- Arduino Uno Mount (thingiverse.com/thing:33327)
- Erik's Arduino Shield (thingiverse.com/download:3919226)
- LightLogo Ring Holder (thingiverse.com/thing:967707)
- Diffuser Cover (thingiverse.com/download:433174)
- Fresnel lens (a.co/a99260S or other lenses like a.co/72ZL9fV or harborfreight.com/rectangle-magnifying-glass-37708.html)

Getting Started

Installing LightLogo

LightLogo comes with an installation guide to help you through the process. Make sure you download the latest version of LightLogo to insure compatibility with your operating system.

1. Install or update Java to the current version for your system: java.com/en/download.

2. Download and install the current FTDI driver for your system: ftdichip.com/Drivers/VCP.htm.

3. Download and decompress the archive containing the current version of LightLogo: playfulinvention.com/lightlogo.

4. Plug in the Arduino to your computer with the USB cable.

5. Open the LightLogo-vm folder and double-click the assembler.jar file to open it.

6. Click on the "asm" button. After a brief pause the console will report back the number of words written.

7. Click on the "download" button to load the LightLogo programming environment onto the Arduino. You will see the transfer (TX) and receive (RX) lights on the Arduino flash and the console will report back an amount of time it took to write the information.
8. Quit the assembler.jar program.
9. Press the reset button on the Arduino.
10. Navigate back to the LightLogo-24 folder that contains the LightLogo.jar file. Double-click LightLogo.jar to open it.
11. Click on the "download" button to transfer the test.txt file to the Arduino.
12. Quit the LightLogo.jar program.
13. Physically disconnect the Arduino from the computer by unplugging the USB cable from the computer.

Connect the NeoPixel Ring to the Arduino

1. You are now ready to double-click the LightLogo.jar file to start programming in the console. But first, you need to set up the NeoPixel ring to communicate with the Arduino.

2. Cut three short lengths of hookup wire, each the same length. Use the wire strippers to remove half an inch of insulation from both ends of the wire. Cut a fourth piece of wire about two inches long and remove half an inch of insulation from each end, too.
3. Insert one wire through the G (ground) hole on the NeoPixel ring. Insert one wire through the Pwr (power) hole. Finally, insert the last wire into the In hole on the NeoPixel ring.
4. Using a fine-tip on your soldering iron, carefully solder the wires to the NeoPixel ring.
5. Solder a 10K resistor to the other end of the In wire. Solder the short piece of wire to the other end of the resistor. The 10K resistor is suggested on the data sheet to protect the NeoPixel ring from voltage spikes, though I and others have built numerous NeoPixel rings for Arduinos that do not include the resistor. Proceed without a resistor at your own risk.

6. Now you will connect the NeoPixel ring to the Arduino. Plug the wire from the NeoPixel Pwr to 5V on the Arduino. Plug the G wire from the NeoPixel to one of the GND connections on the Arduino. Finally, plug the In wire (with the resistor soldered inline) into the Digital (PWM) 2 connection on the Arduino.

7. Connect the Arduino to your computer with the USB cable. A single white LED will illuminate on the NeoPixel ring. If the pixel does not light up, unplug the Arduino and check your soldered connections.

Programming in LightLogo

Now that you have your hardware assembled you are ready to start programming patterns and designs in light using LightLogo, which you previously installed on your Arduino.

1. In the LightLogo-24 folder, double-click the LightLogo.jar file to open it. You will be greeted by LightLogo when it opens. This is the LightLogo console where you can directly interact with the NeoPixel ring and program it to display light patterns.

2. The white pixel that lights up when you first open LightLogo is the turtle. You can program the turtle to move forward or backward along the NeoPixel ring, lighting up pixels and turning them off as it moves. You can even turn off the turtle's light by issuing the command **ht** (for hide turtle).

3. By combining turtle moves and colors, you can program short, repeating or randomized patterns in light.

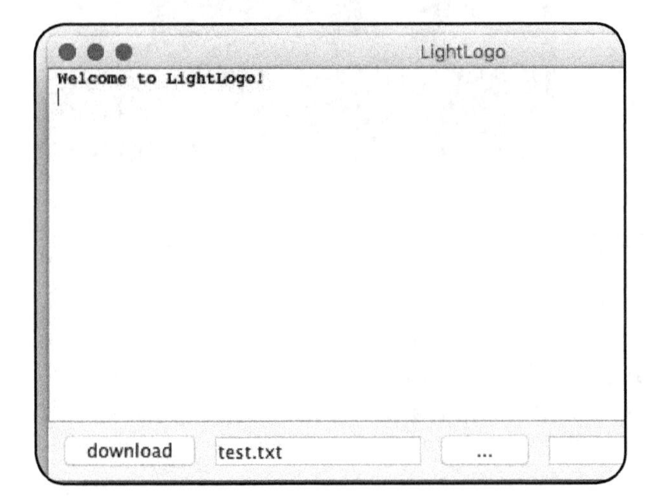

4. The LightLogo Reference manual, which is in the "light docs" folder included with LightLogo, has the entire list of commands available to use in LightLogo as well as short examples of how to use the commands. Be sure to take a look at this manual as it opens up all the features of LightLogo to you. We will start with a couple simple commands:

   fd (forward) - You need to tell the turtle how far forward to move.
   setc (set color) - Sets the turtle's color. The complete list of colors available in LightLogo is included in the LightLogo Reference manual.

5. Type the code at the right in the console, pressing the Return key on your keyboard after every line. The turtle will react instantly each time you press Return.

```
Welcome to LightLogo!
setc red
fd 8
setc green
fd 8
setc blue
fd 8
```

6. As the turtle moves along the ring it lights up the pixels with the colors you define. Since you set the color to red first then move eight pixels, the first eight pixels will be red. When you set the color to green then blue, subsequent moves are also made in those colors. When you finish typing and entering the last command, you will see this pattern displayed on your NeoPixel ring. The turtle is back at the first pixel lit up white.

7. Congratulations! You have started programming patterns in LightLogo.

The Console is for Quick Iteration, But Don't Forget to Save!

The console is a great way to test out ideas; the commands you issue are run on the Arduino but not downloaded, so you can test your ideas and iterate on your designs. For long-term work you want to write your LightLogo procedures in a text editor so you have a record of the work you accomplished that you can return to later. Anything you type and run in the console is not saved nor is it downloaded to the Arduino, so it is important as you start developing more complex work to type the code into a text editor, not the console. Also, you can run the Arduino from a battery, not connected to the computer, to display your light designs if you download the procedure or procedures to the Arduino. Try writing your new procedures in TextEdit on a Mac or Notepad on a Windows PC. If you use TextEdit, you must use TextEdit's preferences to set the document type to plain text, not Rich Text Format. The file you save should end in .txt, not .rtf to be recognized by the console. To download and store the commands you previously wrote in the console on the Arduino, enter these commands in a text file and save it as RGB.txt.

```
to startup
clean
setc red
fd 8
setc green
fd 8
setc blue
fd 8
end
```

LightLogo expects that the procedure that is stored on and runs from the Arduino to be called "startup," so I put the commands I want executed under that procedure. I can call multiple procedures from the startup procedure, something we will explore later in this chapter.

1. Switch back to LightLogo.jar and click the "..." button to navigate to and select the RGB.txt file you saved.
2. Once selected, click the "download" button to transfer the file to the Arduino.

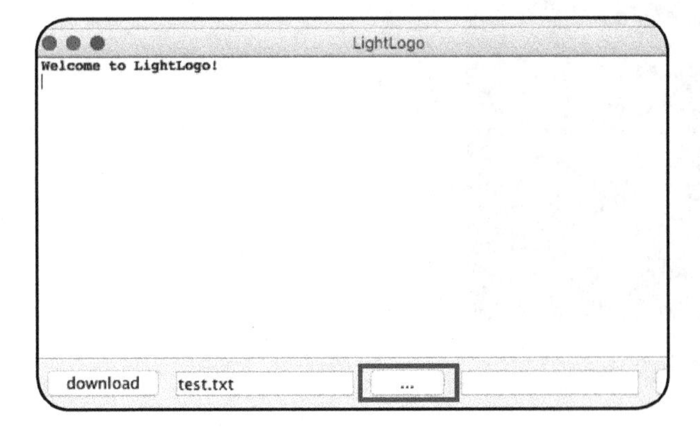

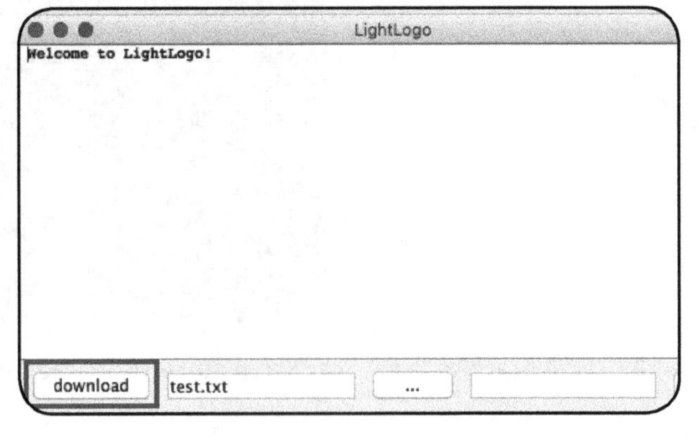

Now your procedure is saved to your computer for later use as well as loaded onto the Arduino. You can power up the Arduino separately from the computer, powered by a 9V battery or a mobile phone charger battery connected by the USB cable. Press the Reset button to run the RGB procedure you created.

Housing Your LightLogo Ring and Arduino

While totally optional, 3D printing an attractive housing for your LightLogo NeoPixel ring and microcontroller adds some protection to the electronics and makes it easy to mount your LightLogo ring in another project.

The Arduino Uno Mount (thingiverse.com/thing:33327) is a 3D printed bracket that works with both the Uno and the Red Board. You attach the microcontroller to the mount with 5mm M3 machine screws or equivalent.

While your 3D printer is still warm, print a copy of Erik Nauman's NeoPixel Arduino Shield (thingiverse.com/download:3919226). This clever shield holds the NeoPixel ring safely in place with room to route the wires to the appropriate places on the Arduino board.

A diffuser is also available to fit on Erik's Shield: thingiverse.com/download:433174. This 3D printed cover makes the pixels a little less bright and a little less distinct, which might be to your liking.

If you wish to use longer wires to connect the NeoPixel ring to the Arduino and mount the ring in a project by itself, you might be interested in my 3D printable NeoPixel ring holder, which has a hole that you can use your choice of hardware to attach the ring to your project: thingiverse.com/thing:967707. I designed this holder on Tinkercad.com, so you can remix the design further to your liking at the link provided on the Thingiverse page.

A great project is to house your LightLogo ring inside a cigar box or pencil box. I had a class of students program narratives in LightLogo (more on that later in this chapter) and house their NeoPixel rings inside pencil boxes on which they collaged images that matched their stories.

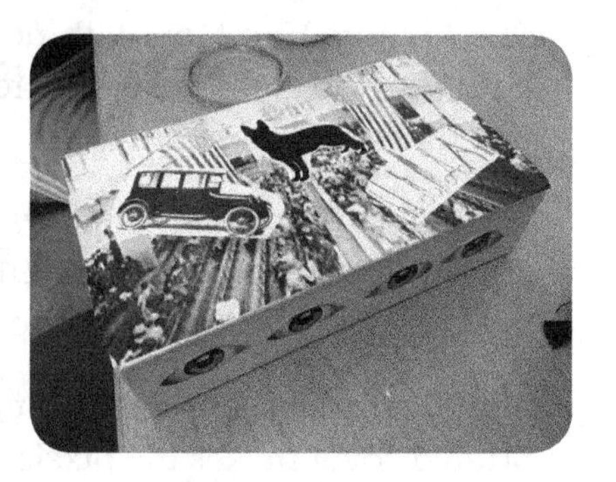

Programming An Animated LightLogo Design

Our first design was static, lighting up the NeoPixel ring in red, green, and blue. Now let's program a procedure that causes the lights on the NeoPixel ring to animate through the pattern of light that we create. Open your text editor and type the following:

```
to every.other
repeat 12 [
setc red
fd 2
setc blue
fd 2]
wait 500
clean
repeat 12 [
setc blue
fd 2
setc red
fd 2 ]
wait 500
end
```

Then above this procedure, type the startup procedure that LightLogo expects:

```
to startup
ht                    ;hides the turtle
loop [
every.other ]
end
```

In the **every.other** procedure we tell the turtle to turn two LEDs red, then the following two LEDs blue, then repeat that eleven more times, filling the NeoPixel ring with this repeating pattern. By telling the turtle to wait 500, we slow down the animation so it is not frantic; feel free to experiment with this number to your liking. Next, we repeat the pattern but reverse the lights: blue first, red second, fill the ring, and wait 500 milliseconds. The startup procedure hides the turtle (**ht**) then calls the **every.other** procedure from a loop, cycling it forever. The semicolon indicates a comment that helps somebody reading the code to understand what the code is doing.

Now that you have created your first animating LightLogo pattern take a deeper look at the LightLogo Reference manual included in the "light docs" folder inside the LightLogo folder you installed. There you can learn all the syntax for LightLogo as well as see some examples to try. By the time you work through this short manual you will be ready to explore LightLogo more in depth.

Storytelling Through Light Designs

One idea to take LightLogo farther is to try storytelling through light patterns. This requires imagination and turning a concrete story into an abstract pattern of lights, but it is easier than you might imagine to plan and program.

When I asked my students to program stories that they heard told about their grandparents, I provided them with a "storyboard" of NeoPixel rings to help them plan the patterns they would program. You can download a copy of the storyboard here: bit.ly/morefunrings.

Students can use their phones to video the light patterns as they narrate their stories or poems.

To illustrate to the students how one might tell a story using light, I programmed a light sequence to a poem much like "Jack and Jill," but starring "Bob and Bill" in LightLogo. Type this procedure into your text editor, save it, and upload it to your LightLogo ring to see it run. You can see a video of it here: youtu.be/VCucHr3xoZ0.

```
to startup
bob
bill
ht
wait 1500
fall1
all red
wait 1500
all yellow
wait 1500
clean
fall2
wait 1000
clean
end

to bob
setc blue
setpos 0      ;sets the position of the turtle on the NeoPixel ring
repeat 23 [
fd 1
wait 100 ]
end

to bill
setc blue
setpos 0
repeat 24 [
fd 1
wait 150 ]
end

to fall1
setpos 23
setc red
repeat 23 [
bk 1
```

```
wait 50 ]
end

to fall2
setpos 24
setc red
repeat 24 [
bk 1
wait 50 ]
end
```

One abstract design I programmed was how the light appeared in various forms throughout the day on Lopez Island, where my family has a house. Here is the code; type it into a text file and try it out yourself.

The first procedure, sunrise, uses variables to change the color of the pixels as well as the brightness as the procedure runs:

```
to sunrise
ht
setbrightness 0
let [num 0]        ;creates a variable called "num" with a value of 0
let [morn 4]
all :num           ;turn on all the pixels with the color set to 0, or red
repeat 20 [
all :num
setbrightness :morn
wait 500
make "num :num + 3 ;increase the "num" variable by 3
make "morn :morn + 1]
end
```

The next procedure, sun, sets up the NeoPixel ring to appear as the sun:

```
to sun
setc yellow
ht
fd 1
end
```

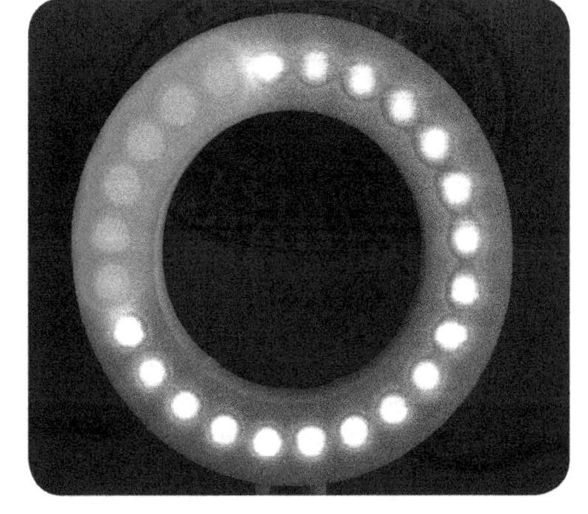

The cloud procedure uses some randomness to generate the size and duration of the clouds:

```
to cloud
setpos random 0 24
setc white
setbrightness 20
st            ;show the turtle
pd            ;put down the pen
repeat random 5 15 [fd 1 wait random 100 250]
ht
end
```

The sunset procedure treats you to a light show, again using variables to change colors and fade the brightness:

```
to sunset
ht
setbrightness 20
all 72
let [num 72]
repeat 24 [
all :num
wait 500
make "num :num - 3]
let [light 20]
repeat 20 [
setbrightness :light
make "light :light - 1
all 0
wait 100]
end
```

Once the sun sets it is time for the star procedure:

```
to star
setpos random 0 24
setc random 40 90
setbrightness random 5 30
ht
Stamp         ;changes the pixel under the turtle to the current color
repeat 4 [setc color + 5]
wait random 1000 5000
end
```

Finally, create a startup procedure at the start of your document to call all the sub-procedures:

```
to startup
loop [
ht
sunrise
setpos 0
st
repeat 24 [sun
wait 50
all yellow
wait 200]
all blue wait random 1000 1750
repeat random 2 10 [cloud
all blue wait random 200 500]
sunset
repeat 18 [star]]
end
```

Using LightLogo with Fresnel and Other Lenses

One project I enjoy tinkering with is projecting LightLogo designs through a Fresnel lens. But why hold the lens with your hand? I built a frame that holds the lens above a base on which to set the LightLogo ring.

My friend helped me design a jig that would hold a Fresnel lens. He drew a shape in Illustrator to laser cut. We started by laser cutting cardboard to check for size and whether the lens would fit. Next, we cut the parts from quarter inch plywood using a laser cutter.

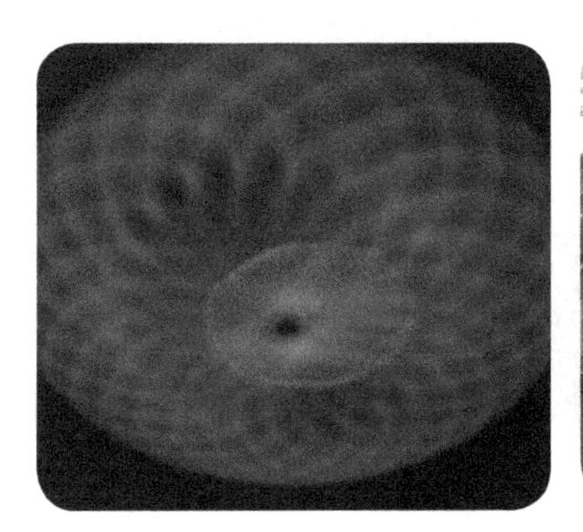

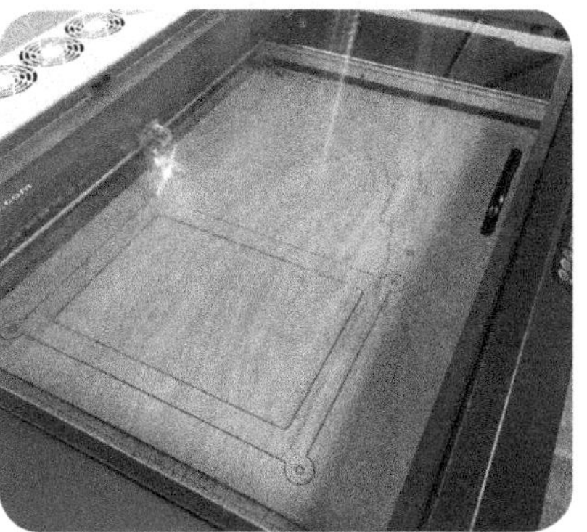

I used threaded rods, washers, and nuts to connect the base and the top. Half inch PVC pipe stabilized the design. Adding additional lenses, like a magnifying glass or smaller Fresnel lenses, provide additional optical effects. There is room enough to have multiple LightLogo rings on the base shining through the lens or lenses.

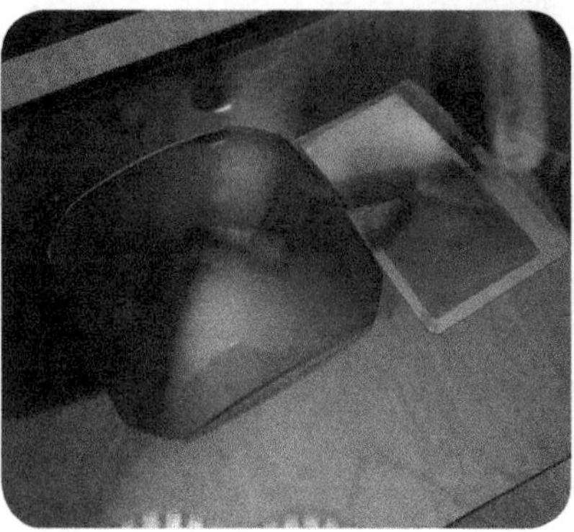

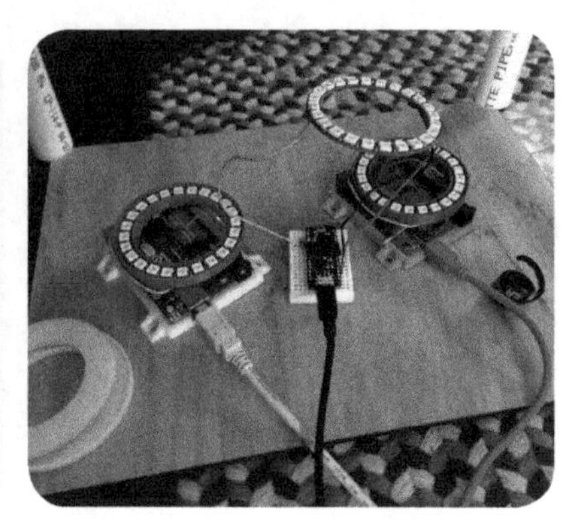

I really enjoy running my Rain procedure on several LightLogo rings through the Fresnel lens at night.

```
to startup
loop [
setc white
setpos 0
repeat 24 [drip wait 25]
reset
setpos 0
setbrightness 20
bigdrop
reset
setbrightness 20
repeat random 25 125
[smalldrop
wait 100]]
end

to bigdrop
all blue
wait 50
clean
end
```

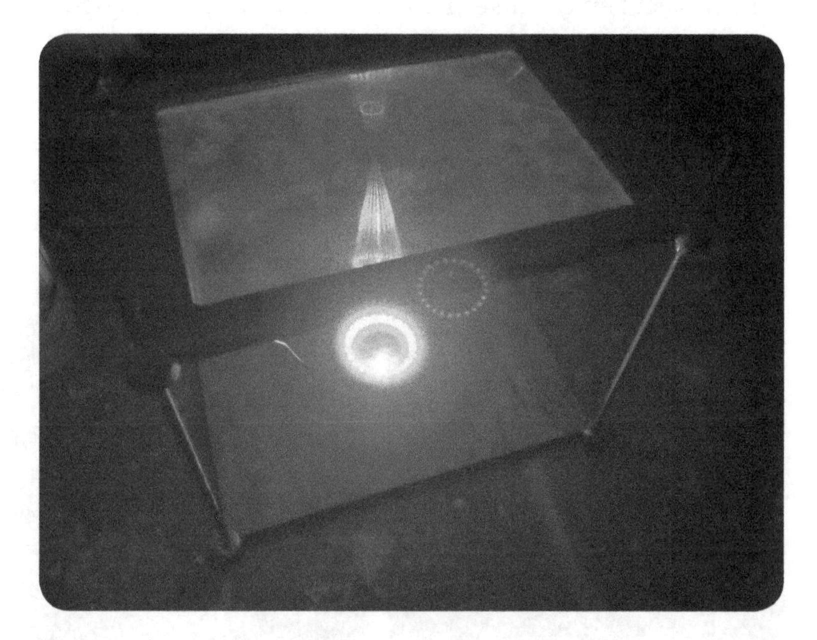

```
to smalldrop
setc blue
setpos random 0 24
stamp
end

to drip
pe ;pen erase
fd 1
end
```

Incorporating LightLogo into Other Toys and Projects

Because of the portability of the LightLogo ring, it is easy to incorporate into projects and play. My son and I played with his wooden blocks one evening, creating monuments that we lit with the LightLogo ring cycling through various colors in simple patterns.

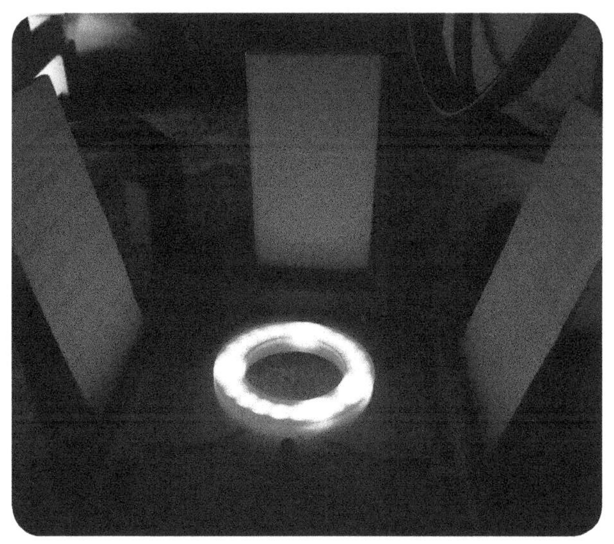

In conclusion, LightLogo is a hard fun microcontroller programming environment. While the floor to entry is a little higher because you use text to program your designs, the syntax is simple yet powerful. By creating small procedures that can add up to create complex patterns you can simplify your debugging. Explore the samples that come with LightLogo, remix them to your liking, and start creating your own procedures.

LEGO Gear Tinkering

Seymour Papert's first "'erector set' project was a crude gear system," he recalls in near Proustian language in the preface to *Mindstorms*. He was adept at imagining gear trains that could turn information around and mesh it to help him construct new knowledge through "making chains of cause and effect." There is little wonder that LEGO includes gears in its Technics, Mindstorms (named after Papert's book), WeDo, EV3, and other programmable sets considering Papert's long association and relationship with the company. LEGO endowed a chair at the MIT Media Lab in 1989 and Papert became the first LEGO Professor of Learning Research (media.mit.edu/people/in-memory/papert) .

LEGO Tinkering is an approach to building and creating with LEGO that originated in online and real life collaborations between Amos Blanton at the LEGO Foundation in Denmark and staff of the Tinkering Studio at The Exploratorium in San Francisco, California. Using the Twitter hashtag #LEGOTinkering, many people, including myself, joined in the hard fun, expanding the LEGO Tinkering mediums from making art to include making music, dance, movement, and more.

When I talk about tinkering I mean working with materials to learn how materials work. LEGO Tinkering is a playful exploration of gears, linkages, beams, and connectors. Once the machine moves it is up to you to decide what it does: can it hold a paintbrush and paint a picture? Can it beat a drum? Can it press a key on a piano repeatedly? You decide. It is likely your machine will not work the first time you power it; LEGO Tinkering is about learning through trial and error, hands-on exploration, and iteration.

This chapter will introduce you to some of the parts used in LEGO Tinkering and help get you started with a few remixable projects. Your interests should dictate where you direct your new LEGO Tinkering skills.

Materials

If you have sets of LEGO Technics and Mindstorms you will likely have the LEGO pieces that you need for the projects in this chapter. However, should you not have riches of LEGO surrounding you, The Tinkering Studio at the Exploratorium Museum in San Francisco offers an à la carte shopping menu of the parts that they recommend for LEGO Tinkering and which I am using in this book, too. You can substitute colors as you see fit for some of the pieces if that makes ordering easy. You can find the list and ordering information at Brickowl, a specialty vendor of new and used LEGO bricks and parts. brickowl.com/wishlist/view/tinkeringstudio/tinkering-with-lego-art-machines. The photos of LEGO components in this chapter are from Brickowl.com.

When I put together my set I was unable to purchase all the pieces on the list but this was not a problem. A hallmark of LEGO Tinkering is flexibility and bricolage, or working with what you have on hand to get the job done. These are parts from the Tinkering Studio's list that I used, which might help you fill out your collection.

Here are the parts to have on hand for the builds in this chapter:

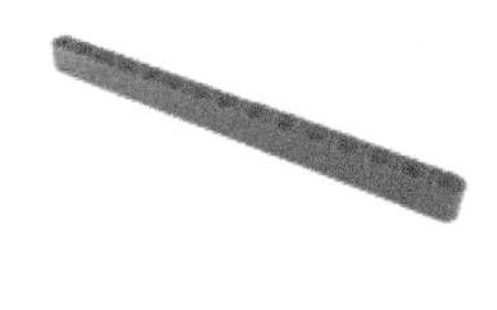

LEGO Red Beam 15 (32278 / 64871)

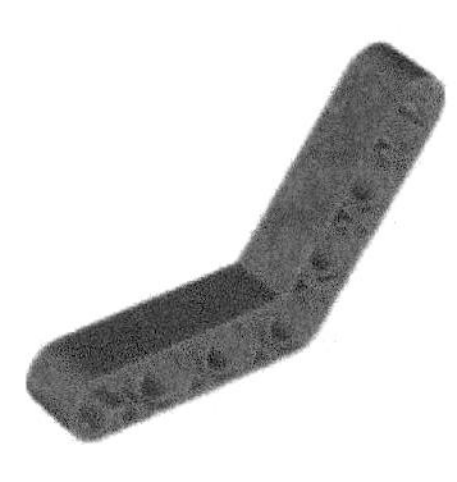

LEGO Beam Bent 53 Degrees, 4 and 4 Holes (32348)

LEGO Axle 9 (60485)

LEGO Beam Bent 53 Degrees, 4 and 6 Holes (6629)

LEGO Gear with 24 Teeth (3648 / 24505)

LEGO Long Pin with Slots and No Friction (32556)

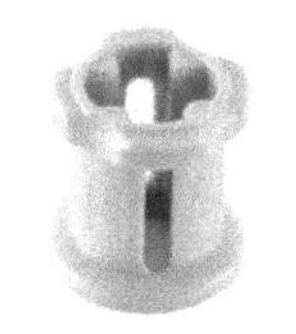

LEGO Bushing (3713 / 6590)

LEGO Pin without Friction Ridges (3673)

LEGO Gear with 40 teeth (3649 / 34432)

LEGO M-Motor (8883)

Power and Hubs

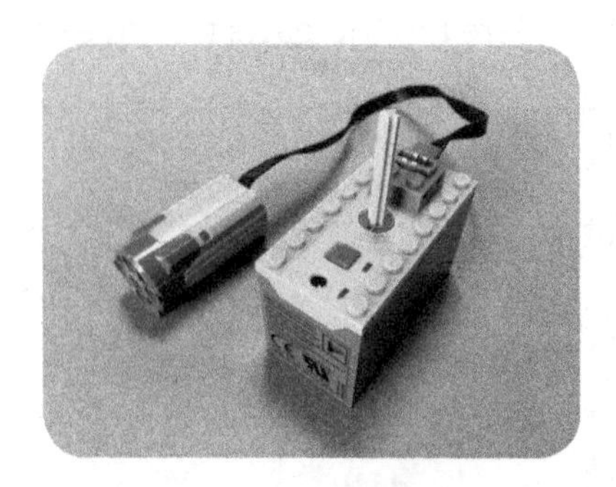

You will also need a power source for your M-Motor. There are several options available from LEGO.

The **LEGO Rechargeable Battery Box** (8788) is great because it is rechargeable and has a dial that allows you to set both the direction of a connected motor as well as the speed of the motor. This Rechargeable Battery Box has a LEGO Axle 7 inserted into the speed and direction control dial to make it easier to turn.

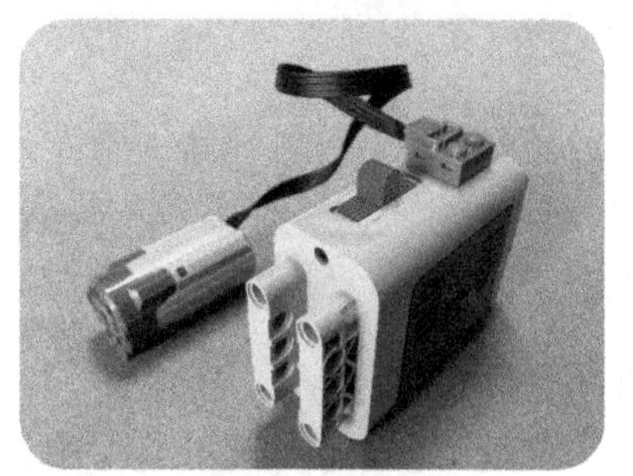

A less expensive alternative is the **LEGO Battery Box** (8881). It uses six AA batteries and does not include a speed setting dial, though you can set the direction of the motor. Motors connected to the Battery Box run at full speed.

The **LEGO WeDo USB Hub** (762615) powers the same M-Motor used by the LEGO Battery Box and Rechargeable Battery Box. Additionally, one of the sensors (the distance sensor is shown connected here) can be connected to the hub at the same time as the motor. See *The Invent to Learn Guide to Fun* for some great art machine projects that combine Scratch, LEGO WeDo motors, and sensors. If you are using the WeDo USB hub your project needs to be tethered to a laptop or desktop computer, unlike the Battery Boxes that can be incorporated into your models.

The **LEGO WeDo 2.0 Hub** connects to a computer or tablet via Bluetooth. The motors and sensors (distance and tilt, like in WeDo) use a different type of connector to plug into the hub, so they are not interchangeable with WeDo or the Battery Boxes. However, the hub and connected motors and sensors do not have to be tethered to a computer or tablet because they communicate wirelessly.

The benefit of both the WeDo and WeDo 2.0 Hubs and Motors is that you can program the motor speed from 1 (very slow) to 10 (fastest). This gives you more control over the machine. However, if you use a Battery Box to power your machine, you can use gears to help you keep your machine under your control (or out of your control, if you prefer).

Gearing Basics

The key to harnessing the amount of speed you want from your LEGO Tinkering machine, whether it runs fast or slow, is using the correct gearing to power the machine's movement. Depending upon the function of your machine you may want to speed up or slow down the motor speed. If you are programming your machine in Scratch or the WeDo software you can easily set the motor speed. If you are using a Battery Box, however, your speed control relies on using gears.

1. Start with a LEGO Beam 15, a LEGO Axle 9 or longer, a LEGO Gear with 24 teeth, and a LEGO Bushing. Assemble them as shown with the axle through the third hole from the right. The LEGO 24 teeth Gear will be the driver gear that is attached to the motor.
2. Add a LEGO Gear with 40 teeth that has a LEGO Axle 9 or longer inserted into the center. Insert the Axle through the Beam so the gears mesh. This is going to be your follower gear, since it follows the movement of the driver gear.
3. Hold the Beam with one hand and turn the LEGO 24 teeth driver Gear. Notice how slowly the LEGO 40 teeth follower Gear turns.

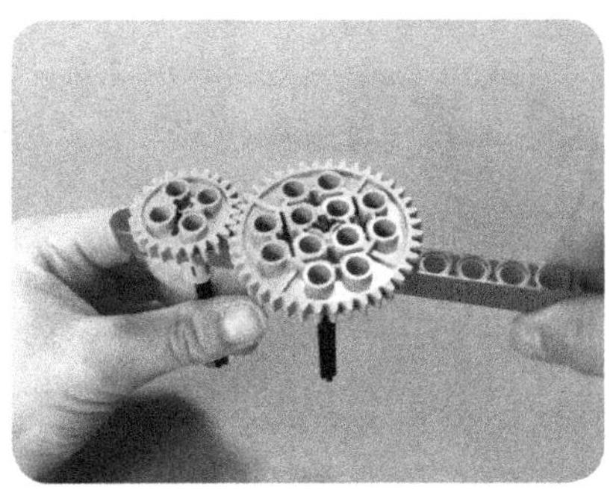

Using a small driver gear and a larger follower is useful to make a machine move more slowly than the motor. However, even though it moves slower, the follower gear has more force as it rotates, called torque. When the driver gear that is powered by a motor is small while the follower gear is large, it is called low gear.

4. Now use your hand to turn the axle attached to the LEGO 40 teeth gear. It is now the driver and the smaller gear is the follower. The tradeoff of speed versus power is reversed. The smaller gear moves faster than the larger driver, but with less power. This is called high gear.

Bullying Machine

This next build demonstrates the difference in torque between a low gear and a high gear. We will add onto the gears you already assembled and build what John Stetson called a "bullying machine" when I met him at the Constructing Modern Knowledge summer institute (constructingmodernknowledge.com) in 2010.

1. Add a 24 teeth Gear to the Axle with the 40 teeth Gear on the other side of the Beam. Add another LEGO 9 Axle and 40 teeth Gear that meshes with the 24 teeth Gear to the Beam as pictured and secure it with another LEGO Bushing.

2. Remove the LEGO Bushing from the 24 teeth Gear and keep it handy. Add a gray LEGO Pin without Friction Ridges to the Beam in the far left hole. Add a tan LEGO Long Pin with Slots and No Friction to the second from the far right hole on the Beam.

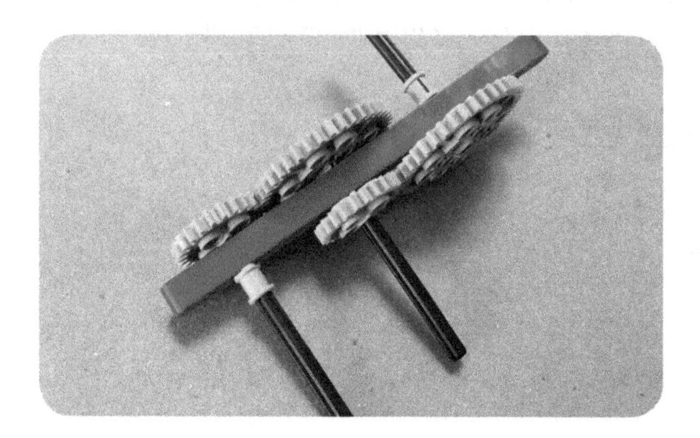

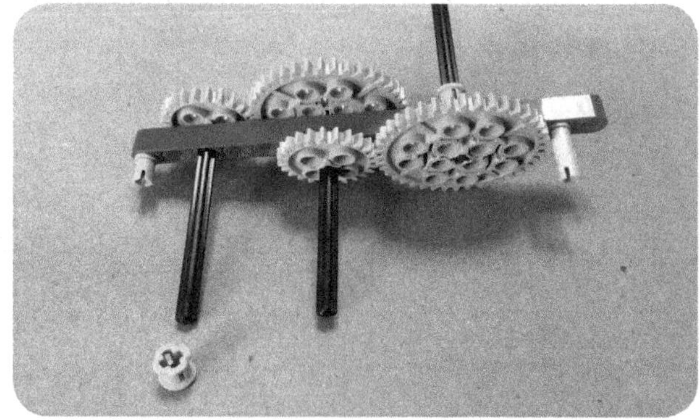

3. Add a Bent Beam with 4 and 4 holes to the gray Pin and LEGO 24 teeth Gear Axle as shown. Turn the machine around and add two grey Pins and a Bent Beam with 4 and 6 Holes to the other side as shown. You are creating handles for people to hold onto when they use the machine.

4. Add another red Beam to the side opposite the first red Beam. Snap the red Beam onto the tan Pin and slide the LEGO Bushing back onto the LEGO 24 teeth Gear Axle. You can also push the Axle with the LEGO 40 teeth Gear on it through the second red Beam to help stabilize the gear in the gear train you built.

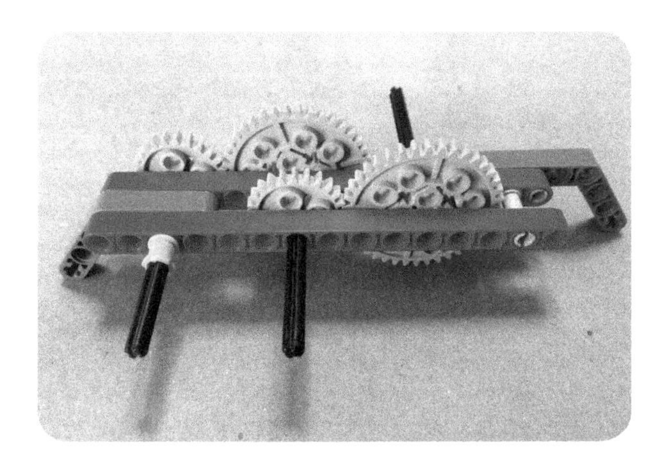

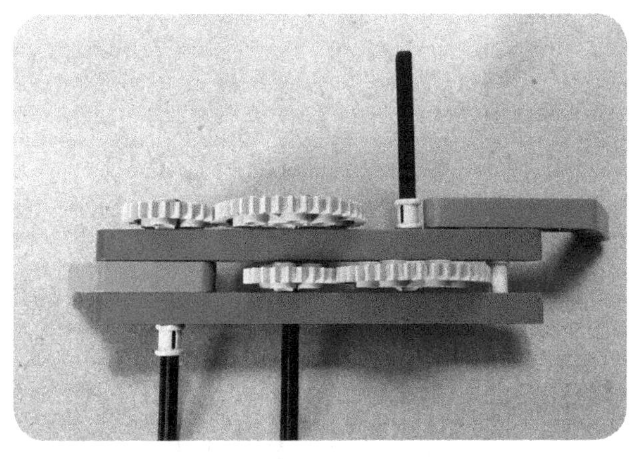

Hold onto one of the handles and have the person you want to "bully" with your machine hold onto the other handle. Next, grip the LEGO Axle to which your gear is attached: if you are the bully you will be holding the Axle with the LEGO 24 teeth Gear, and if you are being bullied you will hold the LEGO Axle with the 40 teeth Gear.

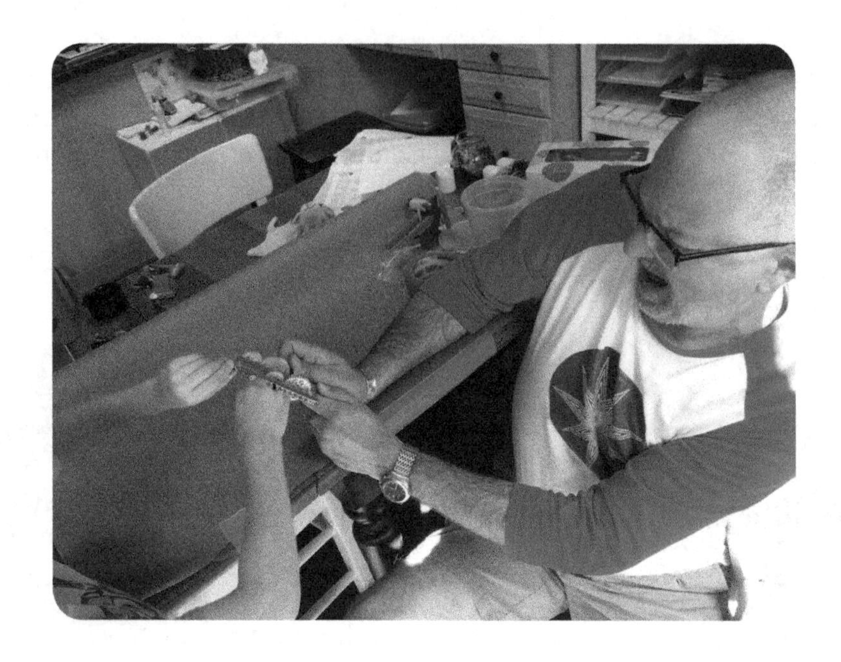

Turn the LEGO Axle with your fingers. The person at the low gear end of the gear train has the mechanical advantage: the person at the high gear end of the gear train cannot stop her or him from turning the gears. Stop turning the LEGO Axle and have the person at the high gear (LEGO 40 teeth Gear) end of the bullying machine start turning the gear train. See if the person at the low gear (LEGO 24 teeth Gear) end of the machine can stop it from turning. The low gear has much higher torque and can immediately stop the gear train.

Idler Gear

Notice that there is an idler gear in the gear train you constructed: the axle in the center (with two gears on either side of the beam) is between the 24 teeth gear and the 40 teeth gear.

With an idler gear in place, when you turn the bullying machine the 24 teeth gear at the front of the gear train and the 40 teeth gear at the end of the gear train rotate in the same direction.

You can use an idler gear whenever you need the output of your gear train to turn in the same direction as the motor and driver gear turn.

A simplified gear train with an idler gear.

Pen Flinger Machine

Let's remix the Bullying Machine and turn it into something less threatening: a Pen Flinger Machine that makes interesting marks on paper. You can experiment with how fast the motor runs and how it affects the pen strokes. While you might be tempted to run it as fast as possible, more nuanced strokes are possible at slower speeds. Using the Rechargeable Battery Box makes it easy to adjust the motor speed. Experiment!

LEGO Technic Pin with Lengthwise Friction Ridges and Center Slots (2780)

LEGO Beam 1 x 2 with Axle Hole and Pin Hole (74695)

LEGO Universal Joint 4 (9244)

LEGO M-Motor (8883)

LEGO Axle 9 (60485)

1. Push the Axle on the 24 teeth Gear until it is flush with the Bushing.
2. Insert two black LEGO Technic Pins into the M-Motor as shown.

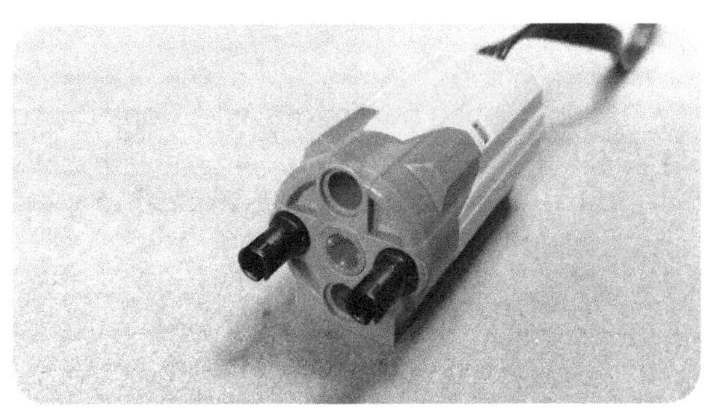

3. Remove the Bushing from the LEGO 24 teeth Gear Axle. Insert the Axle into the M-Motor and snap the LEGO Technic Pins into the red Beam.
4. Attach the Universal Joint to the LEGO 40 teeth gear Axle at the opposite end of the Bullying Machine, on the side opposite the motor.

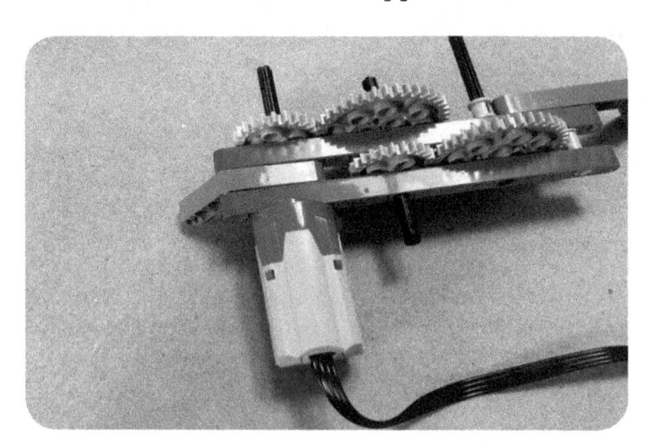

5. Insert the Axle 9 into the Universal Joint.
6. Slide the LEGO Beam 1 x 2 up the Axle until it is about its own width away from the Universal Joint, as shown. Slide the other LEGO Beam 1 x 2 onto the axle.

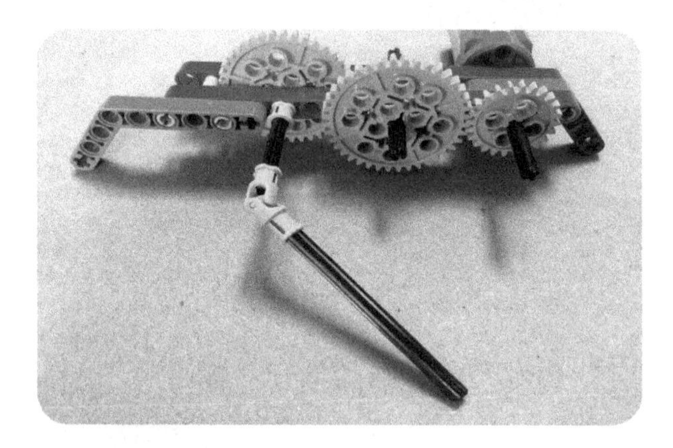

7. Hold the pen next to the Axle. Hook the rubber band around the lower LEGO Beam and Axle and wind it around the Axle and pen. Continue winding until you can hook the other end of the rubber band to the LEGO Beam 1 x 2 at the Universal Joint end of the Axle.

8. Plug in the M-Motor to your choice of Battery Box. I use the Rechargeable Battery Box because I can experiment with the motor speed by turning a dial on the Battery Box. Insert a LEGO 7 Axle into the dial to facilitate setting the M-Motor speed and direction.

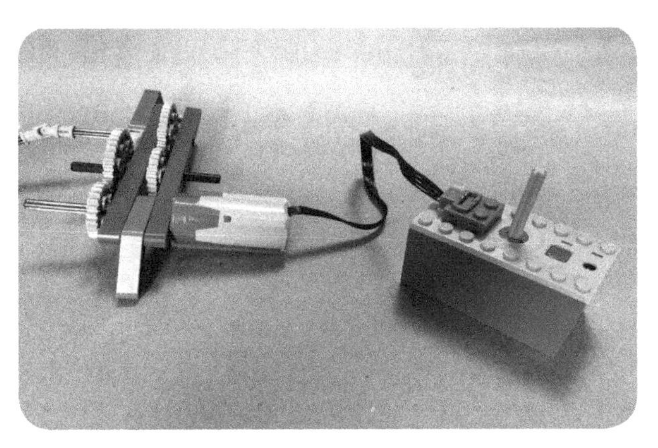

9. Hold the Pen Flinger Machine above the paper as shown. Uncap the pen. Turn on the motor and slowly move the Pen Flinger above the paper, keeping it as parallel to the paper as possible. Experiment with pausing your movement and letting the pen strokes repeat in an area. Some of these strokes were made with the motor running very slowly while others were made with a medium speed.

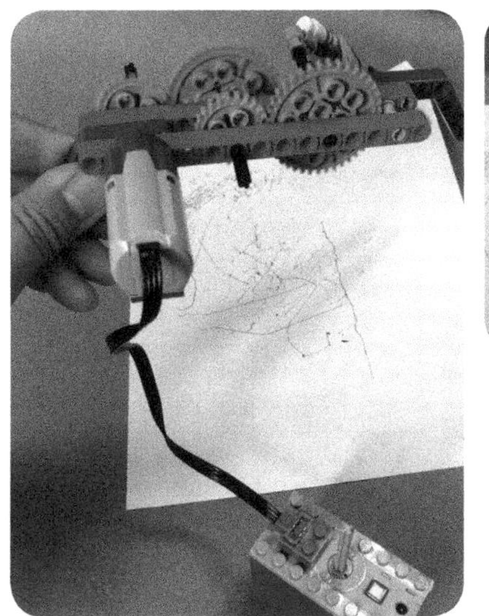

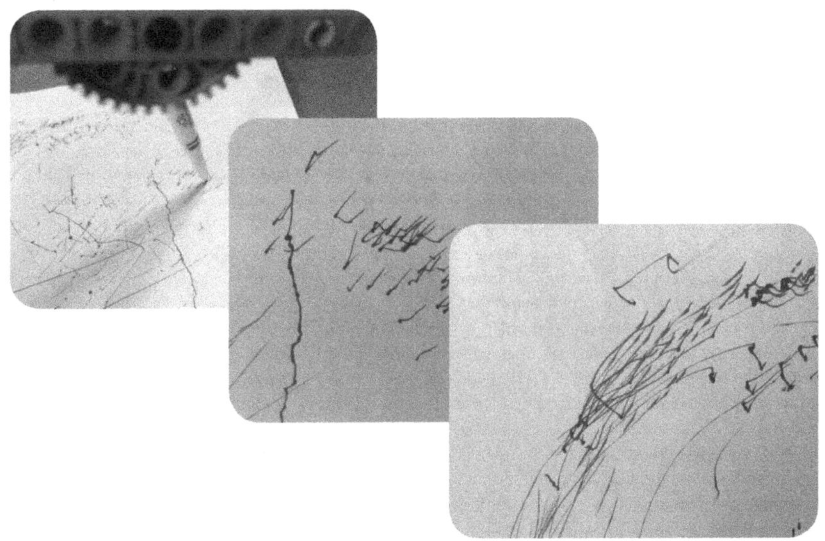

Now it is your turn to remix this design. Give it legs so you do not have to hold it above the table and can concentrate on scooting it around the paper. Add additional pens and rubber bands. The possibilities are only limited by your imagination. Suddenly, you are LEGO Tinkering!

Sound Machine

Next, let's remix one of the Tinkering Studio's Art Machine designs and turn it into a sound machine. The Tinkering Studio provides three base models that are great starting points for LEGO Tinkering here: instructables.com/id/Tinkering-With-LEGO-Art-Machines.

You will need the following LEGO to build your own Sound Machine:

LEGO Technic brick 1 x 16 with Holes (3703)

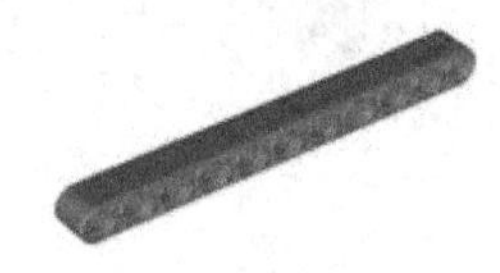

LEGO Beam 11 (32525 / 64290)

LEGO Rechargeable Battery Box (8788)

LEGO Beam 3 (32523)

LEGO M-Motor (8883)

LEGO Beam 1 x 2 with Axle Hole and Pin Hole (74695)

LEGO Pin without Friction Ridges (3673)

LEGO Beam 13 (72747, 72714, 41239)

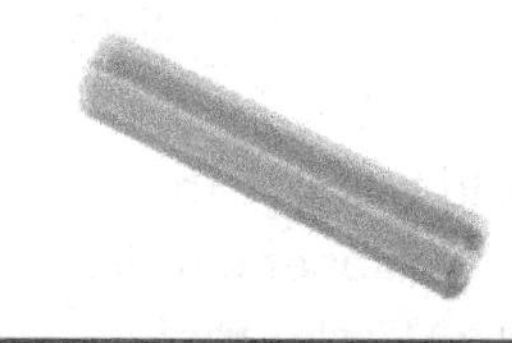

LEGO Axle 3 (4519)

1. Insert a LEGO Axle 7 into the dial on the Rechargeable Battery Box. This makes it easy to adjust the direction and speed of the motor.
2. Connect two 1 x 16 Brick with Holes to the top of the Battery Box as shown.
3. Connect the M-Motor to the Battery Box with the cable. Attach the M-Motor to the 1 x 16 Bricks as shown.

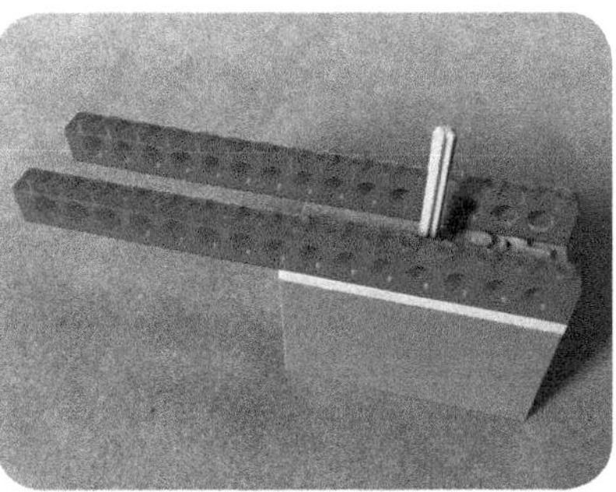
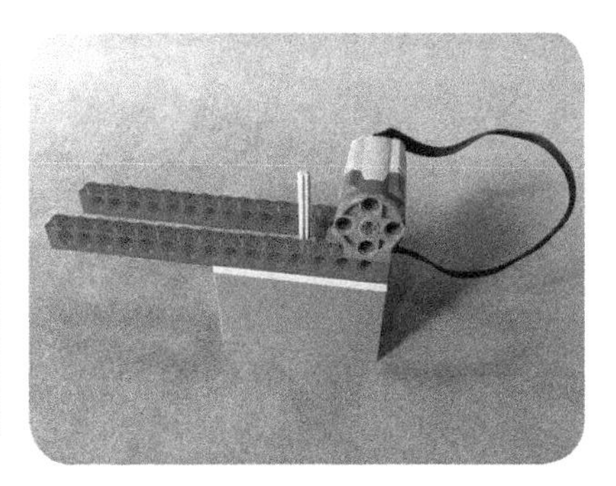

4. Insert a LEGO Pin without Friction Ridges into the top and bottom holes on the face of the M-Motor. Insert a third Pin into the 1 x 16 Brick with holes directly beneath the M-Motor.
5. Insert a LEGO Axle 6 into the orange hub on the M-Motor.
6. Snap a LEGO Beam 11 onto the Pins so the Battery Box continues to sit flat on the table. There will be two empty holes at the top of the Beam.

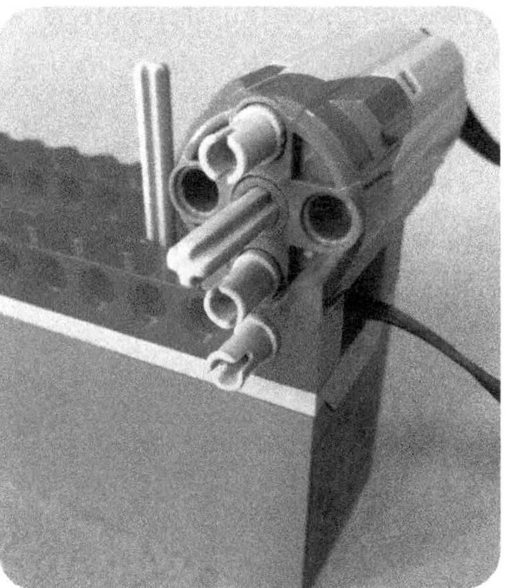

7. Insert a LEGO Pin without Friction Ridges in the third from the bottom hole on the LEGO Beam 11. Attach a LEGO Beam 3 to the Pin in one of the Beam 3's end holes.

8. Slide a LEGO Beam 1 x 2 with Axle Hole and Pin Hole onto the Axle 6 protruding from the M-Motor. Insert a LEGO Pin without Friction Ridges into the other hole on the LEGO Beam 1 x 2.

9. Attach a LEGO Beam 13 to the pins at hole 1 and hole 6.

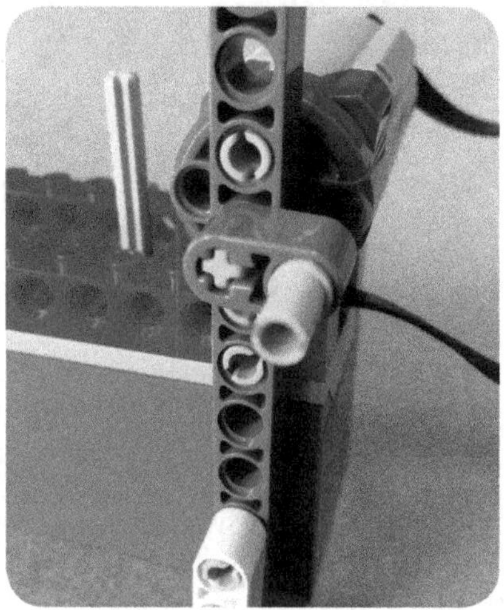

10. Turn on the M-Motor with the button on the Rechargeable Battery Box and observe the machine's movement. The movements of this machine are made possible with linkages. Linkages are assemblies of parts that move together.

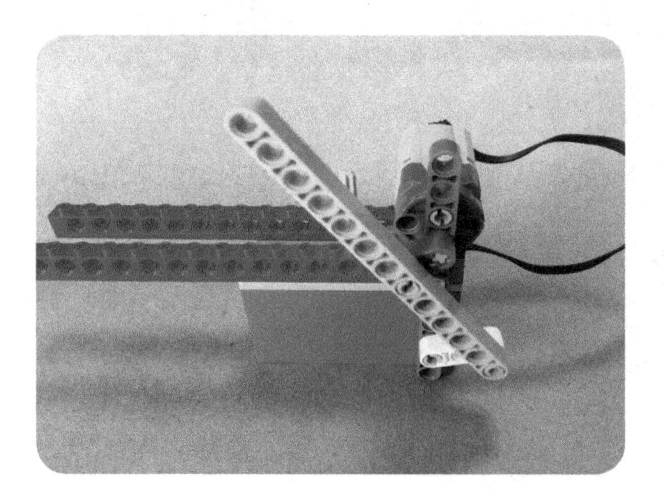

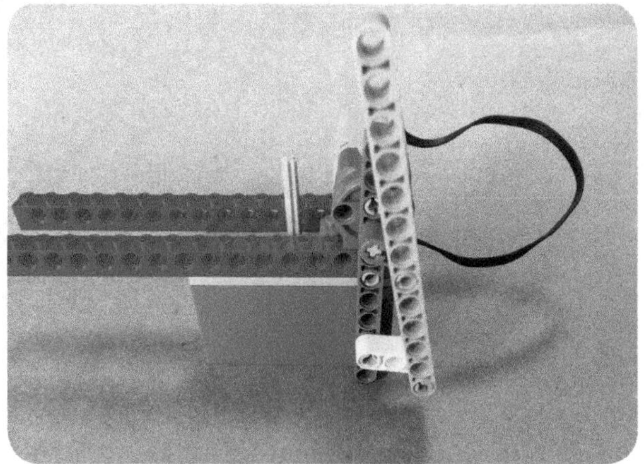

Tinkering with the sound machine

This back and forth movement seemed perfect for banging on a drum. I used a take out soup container as a drum and experimented with different ways to hold the container and let the arm beat against it.

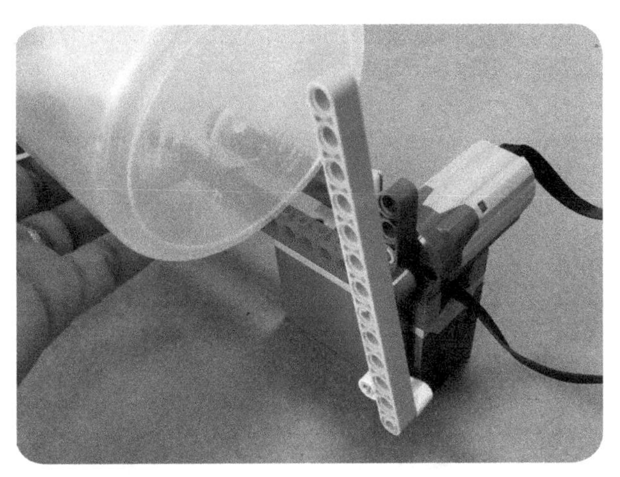

I also taped the container to the 1 x 16 Bricks with the Sound Machine inside the container. The machine moved around lifting and dropping the container as it beat on the inside, making it a movable double-drumming device! (Don't forget—part of LEGO Tinkering is giving your machines cool names.)

This same machine could ring a chime if equipped with a mallet. It could repeatedly press a key on a piano or a computer. If you replace the Battery Box with the WeDo 2.0 hub and motor you could program the machine to make beats that include pauses, varied tempo, and repeating loops.

Linkages

Here are a few more linkages constructed from LEGO to get you tinkering. The peg board on which they are constructed is from a Tinkering Studio design that is customized to work with LEGO pins and beams. It is available to download and laser cut from this link: instructables.com/id/LEGO-Tinkering-Pegboard.

Kinetic Art Machine

This kinetic art machine is a "Lampes Chinoise" construction inspired by Marcel Duchamp's work with rotating, optical illusions.

You can see a video of this Kinetic Art Machine here: youtu.be/4TdLhZzBe3Y.

The Worm Gear meshes with the 24 teeth Gear or the 40 teeth Gear to create a gear train that is slow but which has an incredible amount of torque. The art is attached to the gears on the front of the pegboard, which move on the same axles as gears attached to the gear train behind the pegboard.

LEGO Worm Gear with Old Axle (4716)

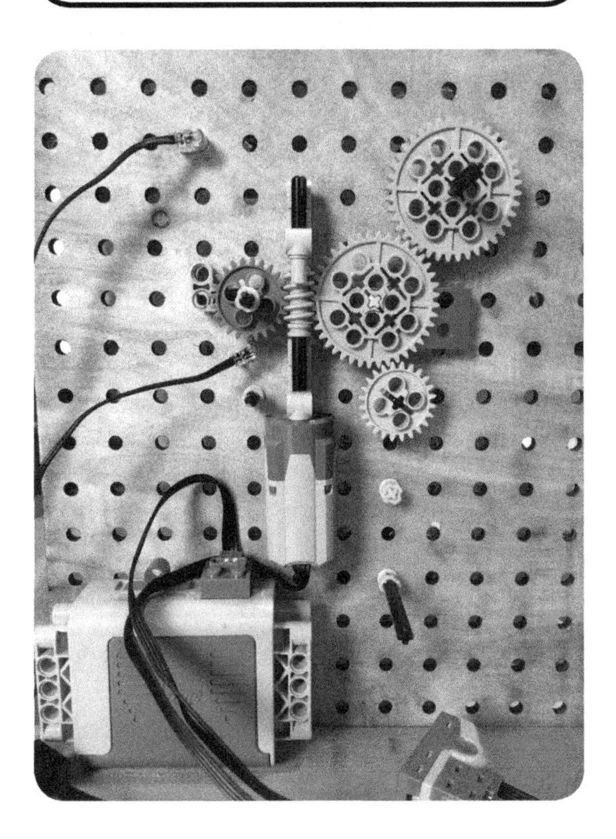

Cam Motion

The Cam included in the Tinkering Studio's LEGO Wishlist is useful for varying the movement of an arm that sits atop the rotating Cam. If you have built the LEGO WeDo Drumming Gorilla model you have used this Cam. A Brick positioned against the Cam will rise and fall repeatedly, making it ideal for percussion, dipping something into a solution, or shaking a rattle. If your machine is positioned on its side, the Cam and Brick can produce a side to side motion, perfect for a sweeping machine, a machine that wipes your nose, or perhaps even a machine that holds a paint brush.

The Cam in this machine makes the 1 x 16 Brick arm with the egg shaped rattle rise and fall, making a sound. The rubber band pulls the 1 x 16 Brick down against the Cam, helping the motion stay in a constant rhythm instead of the Brick bouncing as the Cam spins.

The 3D printed pieces holding the mallet and the pencil are super handy, too. You can find instructions and download the STL files for making them here: instructables.com/id/Tinkering-With-LEGO-Art-Machines.

LEGO Technic Half Beam Cam (6575)

Gearing Calculator

It is difficult to know what combination of gears you need to use in order to slow down or speed up your LEGO Tinkering mechanism. Experimentation is your best teacher, but Sariel's LEGO Gear Calculator (gears.sariel.pl) is an amazing tool when you need an exact calculation. Selecting a driver gear and follower gear from all possible LEGO combinations calculates the gear ratio for you as well how much the speed and torque is increased or decreased, and the follower gear rotation per each revolution of the driver gear.

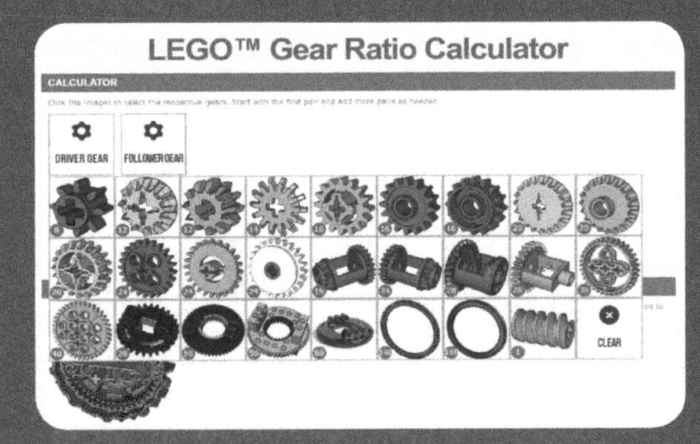

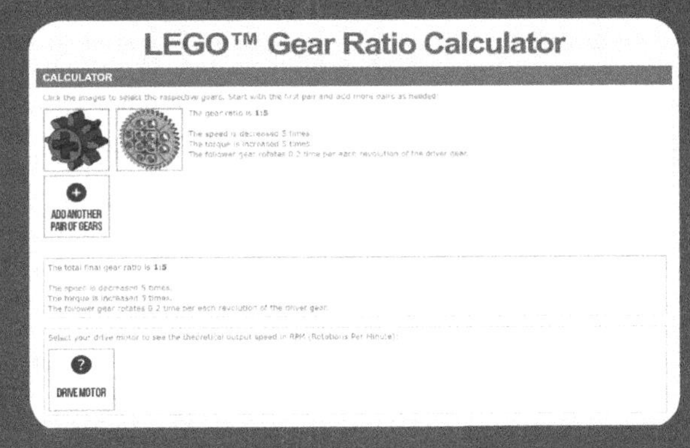

Taking it Further

Now that you have built both an Art Machine and a Sound Machine as well as considered a few other machines built in the LEGO Tinkering spirit, it is your turn to take it further in directions that interest you. Combine gear trains and linkages to create fast or slow, up and down, or side to side movements. Attach brushes, pens, mallets, feathers, or other materials you have to express yourself through your LEGO Tinkering. Share your work on Twitter using the hashtag #LEGOTinkering.

Here are some final resources to consider:

- Sariel's Gear Tutorial is comprehensive (sariel.pl/2009/09/gears-tutorial)
- This LEGO Gear Tutorial explains some of the fundamentals in a clear way (cse.nd.edu/~lego/grp1/Phase2/Report/train.html)
- Weston Middle School's LEGO Mechanisms page is helpful (westonk12engineering.org/robotics/pages/library_of_mechanisms.htm)
- Smith College has a beautifully illustrated tutorial for LEGO Gears (cs.smith.edu/~100e/gears.html)

Keep LEGO Tinkering and having hard fun!

About the Author

Josh is the Middle Division Educational Technologist at The School at Columbia University. He earned a BA in English and Education from Colby College and a MA in Educational Technology from Pepperdine University. He is also a faculty member at the annual Constructing Modern Knowledge institute. Josh taught and worked in technology in both public and private schools for the past 19 years.

He writes books and consults on educational projects that he finds inclusive and disruptive. With Brian Silverman and Erik Nauman, Josh helped create the LogoTurtle robot. During the 2015-16 school year he established an after-school Logo programming club encouraging inner-city students to learn Logo programming, fabrication, electronics, and art. The students programmed algorithmic designs; created t-shirts from their work; learned how to use and maintain a 3D printer; transformed their designs from digital to 3D printed artifacts; assembled the electronics for a Logo-based robot; 3D printed the robot's "body"; and launched in depth mathematical explorations as they created beautiful ink drawings from their procedures.

Additionally, Josh led Horizon National's implementation of a LEGO Community Fund grant to implement LEGO, play-based learning, and STEAM concepts into Horizon's summer curriculum for K-5. Josh's projects have been published in Make Magazine as well as in RED, a Spanish education journal.

An innovative maker educator, his book *The Invent to Learn Guide to Fun* inspires makers of all ages to explore the intersection of crafting and technology through thirteen whimsical, creative projects. The sequel, *The Invent to Learn Guide to More Fun* carries on the hard fun with fourteen all new projects. Josh is interested in culture jamming; making sounds; Logo programming and the LogoTurtle robot; and 3D scanning, modeling, and printing.

You can find him online at joshburker.com, blogging at joshburker.blogspot.com, or on twitter @ joshburker.

The Invent to Learn Guide to Fun

by Josh Burker

The Invent to Learn Guide to Fun features an assortment of insanely clever classroom-tested maker projects for learners of all ages. Josh Burker kicks classroom learning-by-making up a notch with step-by-step instructions, full-color photos, open-ended challenges, and sample code. Learn to paint with light, make your own Operation Game, sew interactive stuffed creatures, build Rube Goldberg machines, design artbots, produce mathematically generated mosaic tiles, program adventure games, and more! Your Makey Makey, LEGO pieces, old computer, recycled junk, and 3D printer will be put to good use in these fun and educational projects. With *The Invent to Learn Guide to Fun* in hand, kids, parents, and teachers are invited to embark on an exciting and fun learning adventure!

The Invent to Learn Guide to Making in the K-3 Classroom: Why, How, and *Wow*!

by Alice Baggett

This full color book packed with photos is a practical guide for primary school educators who want to inspire their students to embrace a tinkering mindset so they can invent fantastic contraptions. Veteran teacher Alice Baggett shares her expertise in how to create hands-on learning experiences for young inventors so students experience the thrilling process of making—complete with epic fails and spectacular discoveries.

The Invent to Learn Guide to 3D Printing in the Classroom: Recipes for Success

by David Thornburg, Norma Thornburg, and Sara Armstrong

This book is an essential guide for educators interested in bringing the amazing world of 3D printing to their classrooms. Learn about the technology, exciting powerful new design software, and even advice for purchasing your first 3D printer. The real power of the book comes from a variety of teacher-tested step-by-step classroom projects. Eighteen fun and challenging projects explore science, technology, engineering, and mathematics, along with forays into the visual arts and design. *The Invent to Learn Guide to 3D Printing in the Classroom* is written in an engaging style by authors with decades of educational technology experience.

Making Science: Reimagining STEM Education in Middle School and Beyond

by Christa Flores

Anthropologist turned science and making teacher Christa Flores shares her classroom tested lessons and resources for learning by making and design in the middle grades and beyond. Richly illustrated with examples of student work, this book offers project ideas, connections to the new Next Generation Science Standards, assessment strategies, and practical tips for educators.

Sylvia's Super-Awesome Project Book: Super-Simple Arduino

by Sylvia (Super-Awesome) Todd

In this superfun book, Sylvia teaches you to understand Arduino microcontroller programming by inventing an adjustable strobe and two digital musical instruments you can play! Along the way, you'll learn a lot about electronics, coding, science, and engineering.

Written and illustrated by a kid, for kids of all ages, Sylvia's whimsical graphics and clever explanations make powerful science, technology, engineering, and math (STEM) concepts accessible and fun.

Meaningful Making: Projects and Inspirations for Fab Labs and Makerspaces

Edited by Paulo Blikstein, Sylvia Libow Martinez, Heather Allen Pang

Project ideas, articles, best practices, and assessment strategies from educators at the forefront of making and hands-on, minds-on education.

CPSIA information can be obtained
at www.ICGtesting.com
Printed in the USA
LVHW06s0902110518
576715LV00001BA/1/P